MOURNING NATURE

# MOURNING Nature

*Hope at the Heart of Ecological Loss and Grief*

EDITED BY ASHLEE CUNSOLO AND KAREN LANDMAN

McGill-Queen's University Press
Montreal & Kingston | London | Chicago

ISBN 978-0-7735-4933-3 (cloth)
ISBN 978-0-7735-4934-0 (paper)
ISBN 978-0-7735-4935-7 (ePDF)
ISBN 978-0-7735-4936-4 (ePUB)

Reprinted 2019, 2022

Legal deposit second quarter 2017
Bibliothèque nationale du Québec

Printed in Canada on acid-free paper that is 100% ancient forest free
(100% post-consumer recycled), processed chlorine free

This book has been published with the help of a grant from the Canadian Federation for the Humanities and Social Sciences, through the Awards to Scholarly Publications Program, using funds provided by the Social Sciences and Humanities Research Council of Canada.

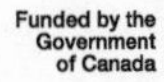
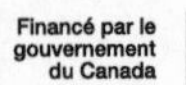

We acknowledge the support of the Canada Council for the Arts.

Nous remercions le Conseil des arts du Canada de son soutien.

Library and Archives Canada Cataloguing in Publication

Mourning nature : hope at the heart of ecological loss and grief / edited by Ashlee Cunsolo and Karen Landman.

Includes bibliographical references and index.
Issued in print and electronic formats.
ISBN 978-0-7735-4933-3 (cloth). – ISBN 978-0-7735-4934-0 (paper). –
ISBN 978-0-7735-4935-7 (ePDF). – ISBN 978-0-7735-4936-4 (ePUB)

1. Environmental degradation – Psychological aspects. 2. Ecological disturbances – Psychological aspects. 3. Global environmental change – Psychological aspects. 4. Grief. 5. Loss (Psychology). I. Cunsolo, Ashlee, 1979–, author, editor II. Landman, Karen, 1954–, editor

GE140.M68 2017 363.7 C2017-900000-4
C2017-900001-2

Set in 10.3/14 Calluna with Calluna Sans and Newslab
Book design & typesetting by Garet Markvoort, zijn digital

To all those who have grieved beyond the human ...

And to those who continue to find hope in the face of ecological loss and degradation ...

May our individual and collective mourning unite, catalyze, characterize, mobilize, and heal.

# Contents

# Figures

# Acknowledgments

A work like this is never possible without the support and commitment of numerous dedicated people, who share a commitment and a passion for seeing a project through over two years. Our first, and sincerest, thanks goes out to the authors in this volume, all of whom lent inspiration, wisdom, passion, patience, and brilliance to this project. Working with the authors has been an honour and privilege, as well as a great pleasure. We deeply thank each one of them for answering the call we first issued in 2012 and responding to the challenge to begin to articulate new ways forward in grieving beyond human bodies, and new ways of thinking about our current ecological struggles.

Second, we are also indebted to the people who helped us at various stages throughout this project, including Raymond Anthony and Brett Buchanan, who provided assessment and feedback on content. Many thanks to the participants and attendees of a panel on mourning with/in the environment organized by Ashlee Cunsolo during the Association of Literature, Environment, and Culture in Canada (ALECC) Conference in Thunder Bay, Ontario (August 2014), including panellists Cheryl Lousley, Brett Buchanan, and Jessica Marion Barr. The ideas and experiences shared through that session provided inspiration when writing the introduction. Our thanks also to the detailed, thoughtful, and useful feedback from the two anonymous reviewers of this collection – your insights and suggestions have improved the works within. We are also very thankful for the time and care that the two anonymous reviewers dedicated to this collection.

We also send much gratitude to Kyla Madden from McGill-Queen's University Press for not only believing in this project from the beginning, but providing helpful information, encouragement, and ideas along the way. Without her, the project would not have proceeded as smoothly!

We extend sincere thanks to Scott Cafarella, who joined us in the final stages to assist with the important tasks of final edits, formatting, cross-referencing, citation lists, indexing, and various other tasks that contributed to the completion of the manuscript.

We also wish to thank Doug Jamieson and his family for the cover photo.

Cape Breton University, the University of Guelph, the Labrador Institute of Memorial University, and the Canada Research Chairs program also deserve thanks for providing the financial and in-kind support required to complete this project.

Finally, we thank our families, friends, and loved ones for putting up with us while we worked on this project, for following us through the highs and the lows, and for being there while we shared the ideas that were emerging in each chapter – all the while reminding us of the importance of unplugging and getting outside to be with more-than-human bodies and places that we love and which ground us deeply. Much love and gratitude.

# Prologue

## *She Was Bereft*

ASHLEE CUNSOLO

The ecologist (in a more than scientific sense) is someone who is touched by this loss in such a way as to mourn the toll of extinction instituted by human exemptionalism and exceptionalism. She is bereft and yet also understands that this feeling, her being touched by irrevocable loss, is itself a matter of realizing the existence of a sense of an ecological and ethical and political community with other species.

MICK SMITH

In 2011, I was bereft. I was at a loss intellectually and emotionally. I felt adrift in waves of sadness, grief, loss, and pain, unanchored from my life, isolated from those around me, and unsure of how to process what I was experiencing. It was a sense of almost abyssal sorrow, without an idea of how to move forward, or how to see beyond the edges, the fringes of these feelings. It was a grief I did not expect or anticipate. Yet it was there waiting for me in the morning when I awoke, there at night as I drifted off, in my dreams and a constant companion throughout the day. And it was related to ecological loss.

My pain was stemming from research that I was doing in partnership with Inuit in Nunatsiavut, Labrador on climate change impacts on Inuit lives, livelihoods, and culture, and the implications for mental and emotional well-being. I was part of a team of Indigenous and non-Indigenous researchers working with the Rigolet Inuit Community Government on a community-led project examining how changes in the weather, sea ice, snow, and plants and animals were connected to Inuit health and wellness. This was a serious concern for people in the community, as changes were happening in the region – happening fast – and people were witnessing their beloved land shift.

For Inuit, the land is everything. It's family. It's kin. It's history. It's heritage. And it's culture. It's a place to refresh, revitalize, and heal. It's a place to experience the blurring of the boundaries between human bodies and non-human bodies. It's a place of reverence and respect. It's a place where relations are forged and strengthened, where children grow, and where families thrive. And it's a place from which wellness flows. Indeed, for a people and a culture relying so intimately and directly on the environment for all aspects of life, livelihoods, and cultural sustainability any shift, however subtle, can have lasting impacts on individuals, families, and communities – physically, mentally, emotionally, and spiritually.[1]

The roots of my bereavement started in March of 2010. I was sitting in the living room of a woman who had agreed to be interviewed as part of the project to share her wisdom and experiences of how things had changed. She was born and raised living on the land, before the forced settlement of Inuit in Labrador, and had a deep connection to and understanding of the rhythms of the land and environment. The land was her people, her cathedral, her life.

We were settled in on a beautiful winter day, sharing tea and stories, and looking out on the frozen bay. We had spent a few hours talking about her life growing up and recent changes she had been witnessing, with the rapidly warming temperatures, the decreasing sea ice, the changing weather, and the alterations to plants, animals, and ecosystems. It was clear that these changes were not only affecting numerous aspects of her life, but were also causing her emotional concern. As we were speaking, I asked her how she felt about all the changes.

She paused ... she looked at me ... and she began to cry.

She cried because of the pain she was feeling as she watched beloved landscapes change. She cried for the impacts it was having on her family and her people. She cried for the sense of loss that was coming as cultural practices, thousands of years old, were being disrupted. And she cried for the changes that had not yet come, but which she was expecting, and what that meant for the land and for Inuit.

Sitting there, in that living room, on that beautiful sunny day in March, sharing that moment, I cried too. I cried because of the emotion

I was witnessing. I cried for the sorrow that was being expressed as a cultural heritage so tied to the land was eroded. I cried because of the depth of her connection to the land, and what she was expressing in such raw and powerful terms. I cried because of my guilt for my complicit involvement in structures contributing to anthropogenic climate change. And I cried for my own pain and sadness at the loss of species and ecosystems from my past, and for what would likely be in my future.

So there we were: a woman from Northern Labrador and a woman from Southern Ontario, united in different forms of ecological grief, sharing our emotions in a way that I had not done before with another person. We were bereft. And while the roots of our ecological grief and experiences were different, and while we were isolated in our personal response, we came together, however momentarily, to share in a loss that was far beyond the human.

The experience was, for me, an altering moment. It was more than catharsis; it was transformative, changing me in ways that I still do not know if I would label "good" or "bad" (indeed, people often want to know this detail from me), but in ways that left me feeling both weak and empowered, full of despair and full of hope. I realized the deep importance of understanding, experiencing, and thinking about ecological grief and mourning, and about the power that comes from having a deep connection to the land and environment in such a way as to leave you completely vulnerable, completely raw.

Throughout this two-year project, our team conducted over eighty in-depth interviews with people in the community. As the interviews progressed, it was increasingly clear that there was an important story to be told about the ways in which climate-related events were negatively disrupting many facets of Inuit mental and emotional health. And so I spent hours listening and re-listening to the audio recordings; I immersed myself in the transcripts; and I spent countless hours talking to friends and colleagues in the community about this work. I was consumed and subsumed by the raw emotion, the passion, and the pain in the transcripts, and was overcome by the sense of loss and grief that I was witnessing. Yet I tried to deny it, for I did not want my own pain to be privileged above that of the people who were brave enough to share their pain with me. I also believed I did not have a right to feel this way, given that those in the North were at the frontlines, and I was but a privileged and honoured witness, sheltered from the daily changes they were experiencing. But their stories also triggered my own ecological grieving – grieving for the pain of others, grieving for the ecological destruction that we, as a species, are causing, grieving for the continued loss, death, and devastation that is to come, grieving for the non-human losses I had already experienced. Throughout all of this, I kept quiet and – as Kristeva described (1989, 4) – pain became for me "the hidden side of my philosophy; its mute sister."

It was as though I were experiencing a deep sense of bereavement, but without the loss of a loved one, which would make the experience more understandable to others (and perhaps to myself as well). There was no one to tell. There was no ritual of mourning to provide comfort or solace. There was no memorializing, no public expression, no community of support that rallied, as when the death of a person occurs. All that was there for me was the pain and the daily realization of human and ecological suffering.

I was consumed.

I could not figure out how to atone for this loss.

I could not figure out how not to be crushed by its very weight.

For researchers, scientists, scholars, artists, and activists working in areas related to extinction, climate change, ocean acidification, deforestation, biodiversity loss, desertification, and large-scale resource extraction (and many others), the work and the knowledge can be existentially and emotionally overwhelming. We know what is happening. We know what can happen. We know the risks. We know the ramifications. And we dedicate every day to trying to create positive change and gain political traction, often to no avail. We surround ourselves with the pain of the world, and try to keep ourselves well enough to move forward in this important work. But it is not easy. In fact, it is often really damn hard to keep going. Amanda di Battista and Andrew Mark share a quote from Ralph Carl Wushke in chapter 9 (232) that speaks directly to this: "For a lot of environmental activists there is a frenetic, desperate sort of activism, in the sense that we've got to act – the earth is going to disappear [and] the planet as we know it is going to be gone – we've got to do something. So there is this intense desperation really ... But one of the things that I've heard when I've talked to people who are professional environmentalists, whose job it is to improve sustainability or recycling, is that there is this sense of despondency, that we're never going to get there, that it's too little too late. So that to me is where a ritual of lament and sitting with the loss so that you actually do the grieving that you need to do [is important]." I know from discussions with fellow colleagues that many of us have wanted to give up, that we wished we did not know and see and experience what we did. But that was only sometimes. Most of the time, we were able to move through the emotions, driven by a sense of urgency and a dedication to our work.

Catriona Sandilands inserts a quote from Andrew Solomon (2001) in chapter 6 in this collection that resonates here: "to be creatures who love, we must be creatures who despair at what we lose" (15). How fitting for a collection on the loss and grief we experience for our more-than-human relations. Indeed, if we truly love what is beyond us, what is around us, what grounds us, and what creates us, we must be prepared to feel the full depth of loss, of grief, of despair, at that loss of non-human bodies. And so, despite my pain, I continued on, because in this work were the teachings and wisdom imbued within Labrador Inuit culture, and a strength and grace in those who are so deeply and so intimately tied to the natural environment. What got me through my bereavement in 2011 was the friendship of those with whom I worked up North, the opportunity to share our experiences together, and the desire to keep forging forward, together, to bring awareness to the changes experienced in the North and the impacts on Northern ecosystems and Inuit homelands.

And so I kept going, motivated by the grief. My grief almost became my happiness, my motivation. No longer debilitated, but inspired by the mourning process, I could keep moving forward, and do so in a way that remembers, recognizes, respects, and pays homage to it.

In 2012, and based on the research that was conducted in Rigolet, we created the Inuit Mental Health and Adaptation to Climate Change (IMHACC) project, an initiative bringing together all five communities in the region to see if the experiences of Rigolet Inuit were shared across the region. Throughout this process, over one hundred in-depth interviews were conducted throughout Nunatsiavut, and the results were the same. People shared deep feelings of loss, sadness, and despair; they expressed anxiety and fear for the future; and they discussed what Glenn Albrecht (2007, S96) has called solastalgia – the "pain or distress caused by the loss of, or inability to derive, solace connected to the negatively perceived state of one's home environment. Solastalgia exists when there is the lived experience of the physical desolation of home."

Many of us on the IMHACC team had been grappling with so many deep emotions throughout the process: grief and loss for the land, a way of life, and a changing culture; anticipatory grieving for what was likely to come with continued climatic and environmental changes; pain as we were confronted repeatedly with others' pain; and, at times, solace, because we were sharing, giving voice to our differing experiences, and coming together to talk about all the mental and emotional pain and distress being witnessed.

But what to do with this despair?

How does one share this pain, and share the stories of those who are connected to this loss?

In late November 2012, during a meeting with the leads from each community, one of the community AngajukKâks (mayor) suggested that we make a film. We had been mobilizing digital media for years as a data-gathering and knowledge-sharing approach that privileged oral history and allowed people to share their own stories, in their own voices (Cunsolo Willox, Harper, and Edge 2012b). And even though none of us had documentary experience, a film was not only a natural progression of this work but a way to introduce people to the wisdom, culture, and experiences of Labrador Inuit as well as share the mental and emotional impacts of climate change within the background of the stunning North Labrador landscape. We shot and co-edited this film over an eighteen-month period, with the final product edited collaboratively with input from 150 people from Nunatsiavut. The final film, titled *Attutaunijuk Nunami*, which translates into *Lament for the Land*, brought together twenty-four people from throughout the region, sharing their experiences and deep emotions, including grief, in response to all the environmental changes. It was released online on 21 September 2014, to coincide with the international day of climate action (to view the film, please visit www.lamentfortheland.ca/film).

The film is gritty and raw in its portrayal and its production, but it seems to have struck a chord with people around the world. We continue to receive requests for screenings, and have heard from people all over – particularly those from the Circumpolar North – saying how moved they were by the film, and how humbled they were by the generosity of spirit shown by the people who were brave enough to stand forward in the film and speak. What has been most interesting, however, is the number of people who have shared that they themselves were experiencing mourning and grief related to environmental loss. We have heard from other Circumpolar Indigenous populations who shared that the Labrador Inuit experience gave voice to what they are experiencing in their own regions as they live at the forefront of global climate change; we have heard from people whose beloved home areas are being desecrated by resource extraction and development; we have heard about the loss and despair experienced over species disappearance in a region; and we have heard about the trauma felt when witnessing deforestation and clear-cutting. All these people, in all these places, united by a deep sense of connection to the natural world, are stepping forward to collectively share our grief in response to the public sharing of others.

In this, we take heart.

There are many of us who are not only sharing these emotions, but are ready to talk about them, and to learn from others, in order to move forward under the weight of the enormity of what we face. What also struck us was how emotional people were during the public screenings of this film. So many viewers were overcome by emotion. Some shed tears. Some had to leave. Some could not talk. Some expressed deep guilt and shame that they did not realize people in the North were experiencing these changes, and were overwhelmed by the realization of their own complicit contribution to large-scale climatic events. And many were unable to process their full responses during the viewing, choosing instead to contact us days, weeks, or months after the viewing. And yet in the midst of those reactions there was also so much collective hope that was shared, as people expressed their ecologically based grief with others and realized that, in this, they were not alone.

We also realized, after, that our film was in a way responding to Robert Nixon's (2011) call in *Slow Violence and the Environmentalism of the Poor*, when he asks readers: "how can we convert into image and narrative the disasters that are slow moving and long in the making, disasters that are anonymous and that star nobody ... How can we turn the long emergencies of slow violence into stories dramatic enough to rouse public sentiment and warrant political intervention?" (3). Indeed, I feel – and I believe this collection makes clear – that it is in our stories, in our writings, in our philosophies, in our sharings, and in our songs, protests, art, and other expressions that we also take up this challenge and confront those slow-moving environmental traumas that Nixon labels slow violence, including climate change, extinction, deforestation, and the acidifying of the world's oceans.

In this, I take heart.

In 2014, Jo Confino published a piece in the *Guardian*, where he posed two questions that I think of often: "Why aren't we on the floor doubled up in pain at our capacity for industrial scale genocide of the world's species?" and "Is it possible to hold all the grief in the world and not get crushed by it?" He asks these questions, "because our failure to deal with the collective and individual pain generated as a result of our destructive economic system is blocking us from reaching out for the solutions that can help us find another direction."

I think in many ways, this is what the collection is about and what motivated its beginnings. We are all, through different approaches, seeking, grasping, reaching, moving towards another way, a new direction forward. One that does not see us crushed by loss only – a pathway that allows for, and then moves beyond, this "doubling up" in pain. One that allows us to unite our scientific data, our research, our theories, and our thinkings together with our deep emotions and our ecological grief.

The question is, then: how can we embrace the pain to learn from it, express our grief, and then transcend? This is no easy task. But the authors in this collection each take up this challenge in their own way, and with great care and creativity.

I have loved this collection. I have loved reading each of the chapters within with great admiration, with a deep appreciation for the intellectual and emotional commitment given over by the authors. But there were times when I could not face it. Not because of the task itself, but because of the ways in which my own emotions and ecological grief were intertwined with this work. There is pain within these chapters, and the work in this collection stands as a reminder, at times lurking at the edges of my subconscious, at times directly in my line of sight, of my own grief that is tied up in ecological loss and processes, of my own sense of helplessness and despair. But also, within – and from where I find solace – reflections of my own sense of hope for change.

To write this opening, then, is to be faced with – to be forced to come to terms with – deep loss and deep pain that resides within, and which surfaces unexpectedly, and occupies the places and spaces that open up when I allow myself to feel – to *really feel* – this loss. But to write is also a way to confront the loss, to embrace it, to examine it from all facets, to try to understand it, to plumb the depths so that I may understand the terrain well enough to journey out and beyond. And to make this journey with the many others who are also thinking and feeling with and through the work of mourning within contexts beyond human bodies.

I think, in many ways, this collection is full of authors who are confronting their own pain and their sense of urgency about what is happening to the more-than-human worlds, and adding their intellectual strengths and rigour to better articulating why ecological grief and mourning need to be widely discussed, analyzed, and considered. This collection, then, is as much an ecological elegy and eulogy as it is a philosophical and intellectual contribution. It's also about loss of place. A loss that is both literal in that the landscapes and ecosystems around us are changing, and psycho-social because on a deeper level it can feel that we are losing our place in the world, that we do not know how to stem the destruction, heal the wounds, make amends, atone, and reconcile. But we do know that we have to act. And we do know that these challenges we are facing, as individuals and as a species, will require different responses. Perhaps, then, ecological grief and mourning may be ways to inform and motivate understanding, action, and change. And if that is so then, as Morton (2007, 185) writes, "now is a time for grief to persist, to ring throughout the world."

*Unama'ki/Cape Breton, Nova Scotia*

NOTES

1 To read more about the research that came out of this work, to learn from the Labrador Inuit, and to read detailed quotations and climate science, please see Cunsolo Willox et al. 2012a; Cunsolo Willox, Harper, Edge, et al. 2013a; Cunsolo Willox, Harper, Ford, et al. 2013b. In addition, there are a number of digital stories, created by Labrador Inuit, available at https://www.youtube.com/user/uKautsiga.

## REFERENCES

Albrecht, Glenn, Gina-Maree Sartore, Linda Connor, Nick Higginbotham, Sonia Freeman, Brian Kelly, Helen Stain, and Georgia Pollard. 2007. "Solastalgia: The Distress Caused by Environmental Change." *Australasian Psychiatry: Bulletin of Royal Australian and New Zealand College of Psychiatrists* 15, no. 1: S95–8. doi:10.1080/10398560701701288.

Confino, Jo. 2014. "Grieving Could Offer a Pathway out of a Destructive Economic System." *Guardian Sustainable Business*, 2 October. http://www.theguardian.com/sustainable-business/2014/oct/02/grieving-pathway-destructive-economic-system. Accessed 8 October 2015.

Cunsolo Willox, Ashlee, Sherilee L. Harper, James D. Ford, Karen Landman, Karen Houle, and Victoria L. Edge. 2012a. "From This Place and of This Place: Climate Change, Sense of Place, and Health in Nunatsiavut, Canada." *Social Science and Medicine* 75, no. 3: 538–47.

Cunsolo Willox, Ashlee, Sherilee L. Harper, and Victoria L. Edge. 2012b. "Storytelling in a Digital Age: Digital Storytelling as an Emerging Narrative Method for Preserving and Promoting Indigenous Oral Wisdom." *Qualitative Research* 13, no. 2: 127–47. doi:10.1177/1468794112446105.

Cunsolo Willox, Ashlee, Sherilee L. Harper, Victoria L. Edge, Karen Landman, Karen Houle, and James D. Ford. 2013a. "The Land Enriches the Soul: On Climatic and Environmental Change, Affect, and Emotional Health and Well-Being in Rigolet, Nunatsiavut, Canada." *Emotion, Space and Society* 6, no. 1: 14–24.

Cunsolo Willox, Ashlee, Sherilee L. Harper, James D. Ford, Victoria L. Edge, Karen Landman, Karen Houle, Sarah Blake, and Charlotte Wolfrey. 2013b. "Climate Change and Mental Health: An Exploratory Case Study from Rigolet, Nunatsiavut, Canada." *Climatic Change* 121, no. 2: 255–70.

Kristeva, Julia. 1989. *Black Sun: Depression and Melancholia*. Translated by Leon S. Roudiez. New York: Columbia University Press.

Morton, Timothy. 2007. *Ecology without Nature: Rethinking Environmental Aesthetics*. Cambridge, MA: Harvard University Press.

Nixon, Rob. 2011. *Slow Violence and the Environmentalism of the Poor.* Cambridge, MA: Harvard University Press.

Smith, Mick. 2013. "Ecological Community, the Sense of the World, and Senseless Extinction." *Environmental Humanities* 2: 21.

Solomon, Andrew. 2001. *The Noonday Demon: An Atlas of Depression*. New York: Scribner.

# MOURNING NATURE

# Introduction

## *To Mourn beyond the Human*

ASHLEE CUNSOLO AND KAREN LANDMAN

The dead no longer walk or appear among the living as they once did; the worldly possibilities resident in their singular being are lost or dispersed; their significance is exposed to the whims of memory and history; the experiential opening onto the world that was theirs, and theirs alone, is extinguished; their constitutive, yet hardly fathomed, roles in the wider community fall empty. This is not to say that life will not go on, that different possibilities will not arise, that the world will cease to be made meaningful to, or experienced by, others, or that community necessarily collapses, but to notice an irredeemable loss, a loss that even eternity cannot rectify.

MICK SMITH

### Introduction: Mourning with/in Nature

This collection is about loss. It is about grief. And it is about mourning. But it is also about understanding absence, sense of place, and the spectral haunting that comes from more-than-human loss. It is about elegies and auguries, melancholy and transformation, and about different ways of knowing and being in the world that stretch beyond solely human bodies into the sensuous experiences of the more-than-human world(s). It is about traces and memories and awareness beyond the human. It is about decentring subjectivities, healing environmental grief, and living connectivity and interdependency. It is about mourning that resists the artificial separation between bodies that can and cannot be mourned. It is about asking what counts as a mournable body (and what does not), and it is about thinking beyond the human to extend the work of mourning to non-humans to think about other possible futures, other possible mournings. It is about recognizing our shared vulnerabilities to human and non-human bodies, and embracing our complicity in the death of these other bodies – however painful that process may be. It is about finding ways to move beyond "human exemptionalism and exceptionalism that allow little or no space for considering other species as parts of the *same* community as ourselves at all" (Smith 2013, 21). And it is about asking – and seriously considering – the question Judith Butler poses regarding what, then, may be gained from "tarrying with grief, from remaining exposed to its unbearability and not endeavoring to seek a resolution for grief" (2004, 30)?

The idea for this collection grew from our own personal experiences with deep and profound moments of grief for non-human entities and degrading landscapes and ecosystems, and our own witnessing of the ways in which others were beginning to share their own personal experiences with ecologically related grief. We were inspired by the expressions of mourning for the more-than-human world(s) revealed through art, poetry, prose, and music, and by the stories that we were privileged to hear from our friends, family, and colleagues. And, as you have read in the prologue, we were particularly inspired by the voices and narratives shared by Inuit in Nunatsiavut, Labrador, about the ways in which shifts stemming from climate change were tied up with a sense of loss – of place, culture, livelihoods, and beloved environments – and how thinking with and through the work of mourning could change ethical and political discussions of climate change (see chapter 7).

Nevertheless we weren't sure that others would find this topic compelling so we decided to release a general call for papers on this topic in late 2012. While we imagined there might be some interest, we were completely inspired, excited, humbled, and motivated by the rapid and overwhelming response we received. From over sixty abstracts submitted we narrowed it down to twenty-five for full-text submission and peer review. After an internal and external review process, we selected the eleven chapters in this collection for their diversity and complementarity, and for their abilities to illustrate the complexities and nuances of mourning beyond the human by juxtaposing multiple perspectives, experiences, understandings, and representations of ecological loss and mourning.

Yet despite these experiences and stories shared by the authors in this collection, and by people everywhere, and despite the fact that many people are intimately familiar with grief processes that move beyond human bodies, there is a surprising lack of discussion around mourning related to environmental loss or dispossession in broader discursive frameworks or public dialogues. This caused us to wonder about the mechanisms and process that were and are keeping such discussions silent or at least preventing them from being shared in broader contexts. Indeed, given our own experiences and the experiences of others, what is puzzling to us is why this conceptual, theoretical, and philosophical leap of extending mourning beyond the human body and leveraging the political potential of mourning has not been made, and why it still lies in the margins of contemporary politics, philosophy, and public narratives.

Perhaps it is because grief work is so deeply painful and, at times, debilitating?

This is undeniably hard work. Perhaps it is because, if we were to truly open up to the full extent of mourning for not only the human lives that we derealize and invisibilize but also our non-human bodies, we simply could not bear the true weight of what that would entail?

Indeed, as you read through the chapters you will see the signs and the traces of deep pain and loss, and see the transformative effects that such loss has had. For many of the authors (we editors included), even writing and preparing this collection was an act of grieving tied up in the labours of mourning. Many of us experienced our own grief responses, our own grief work, as we tried to share the ideas in this collection in narrative form, and worked through our own intellectual and affective experiences with, and commitment to, this work and these labours. We experienced death and loss, dealt with deep pain and personal grief. At times we found solace in writing on these topics, and at other times we could not face the task of thinking through grief while simultaneously experiencing our own. We also found hope and inspiration – hope that there is potential to mobilize pain and sorrow for positive action and change; and inspiration, because we were encountering so many others who were not only interested in but in fact already doing this work.

We are living in a time of great environmental change, where the pace and scope of these changes are rapidly increasing. Climate change, deforestation, ocean acidification, melting polar ice caps, and desertification are degrading and altering places and ecosystems, with serious implications for the multitude of bodies that inhabit them – human and more-than-human. We are also living in a time when it seems as though the lived experiences of those who are dealing with environmental alteration and degradation have outstripped our ability to think with and about these changes. "The awareness of human culpability at a global scale," writes Glenn Albrecht in chapter 11 (296), "is also a relatively new experience in the history of human mourning. Many humans now understand that 'we' are often the primary agents causing disaster to impinge upon ourselves. Environmental pollution, global warming, bad urban planning, and poor engineering are now major anthropogenic causes of human misery and death. In the Anthropocene, there is no longer mystery attached to a great deal of disaster and misfortune since, to a very large extent, there is an element of self-imposed vulnerability to what are euphemistically called 'natural disasters.'" How do we even register the grief, and how do we accept what we have done and for which we are responsible? How does one atone for this loss, to reconcile one's role, and to make sense enough of the grief and all the grievable lives to move forward?

We are entering a time when ecologically based mourning seems likely to occupy more and more of our experience. And yet it seems that we are also entering a time of great denial and avoidance of this type of work because of what it means and what it will entail if we were to truly embrace and open ourselves to these changes and this mourning, and to the understanding of our individual and collective responsibility. We are all implicated in this loss; indeed, our very lives and existence are predicated on the deaths of other bodies that have come before use – human and more-than-human – and on the promise of future deaths. How do we even begin to think and feel *that*?

Given the immense challenges our planet currently faces, we need a mechanism for moving into new terrains of thought that may provide avenues for thinking with and through environmental challenges, for encouraging action, and for potentially cultivating new emotions in fruitful ways. With this in mind, the primary concerns of this collection lie in trying to find new environmental thought and action premised on the work and labours of mourning, and to ask what – politically, ethically, and theoretically – can the work of mourning do if extended beyond the human, and what might this tell us about ecological ethics, politics, and action? How does grief help us live better with others? How do we understand, think, and feel the meaning of all the ecological loss? How do we make sense of what is happening and what is already set in motion environmentally? How do we come to terms with the knowledge that there are species, ecosystems, and cultures that will not survive the environmental degradation and effects of increasing climate change? How can we understand – and cultivate – our shared vulnerability to environmental loss? How do we both think and be with more-than-humans? How can we resist the increasing commodification of bodies – our own and others – through mourning? How do we move beyond theory to action, beyond intellectual exercise to something that is action-based while also being grounded in theoretical strength and understanding? How do we signify our connections to the more-than-human world(s) in new ways – ways that better reflect our experiences, our affects, our feelings, our emotions? How do we make visible our connections to the more-than-humans that are hidden, silenced, marginalized, and which we are not taught to see? And how do we understand – truly understand – the meaning of it all? What is entailed in the ecological work of mourning? What can our labours offer the more-than-human world(s), of which we are also a part? And will this bring us closer to a different form of ecological ethics?

These are some of the many complex questions that this collection attempts to address, puzzle through, and gain traction with to allow our theoretical work to catch up to the lived experiences of environmentally based grief and mourning – the grief and mourning that people around the world are already experiencing. Given the scale, scope, and speed at which we are facing environmental degradation and alteration, we face even more impetus than ever to think differently about our relationships to and with the more-than-humans, and to move beyond those humanist approaches to mourning that privilege human bodies (and not even all human bodies). Indeed, this collection grew from our own sense of intellectual desperation to try to find ways to understand, to think through, to think with the ecological loss that we were both witnessing and experiencing.

This is, for us, more than a theoretical exercise – it is deeply personal and urgent.

So what can we do? And what, exactly, is entailed in the work and labours of mourning?

## The Work and Labours of Mourning[1]

By virtue of having a body, we are all vulnerable to loss. We are vulnerable to losing others with whom we are connected, and we are vulnerable to losing ourselves. Our very existence is predicated on those who have come before us, those we will survive, and those who will survive us. From the moment of our birth we are confronted by the inevitable end of ourselves and of others and also, therefore, by the eventual loss and grief that accompanies these partings. We are, already, preparing to say goodbye and are, already, caught up in an existence in which mourning cannot be avoided (Derrida 1996). And this is where the "work" of mourning lies: in understanding our shared vulnerability and our shared finiteness; embracing the responsibility of the labours of grief that this entails; and understanding that all we have to give to the dead, to what we have lost, is our own living and our own acts of mourning and remembrance (Brault and Naas 2001). Mourning, then, is a call to responsibility to engage with what was lost and, in this light, becomes work to which we must always attend (Derrida 1996).

In his seminal 1917 essay "Mourning and Melancholia," Sigmund Freud outlined an understanding of mourning as long, hard, laborious work that was never truly concluded. For Freud, mourning is a process full of often-uncontrollable emotional and corporeal responses such as grief, pain, anguish, sadness, devastation, denial, and corporeal affects that emerge from the shock of losing something or someone that was loved, valued, and important (Dubose 1997). Mourning is also a work that cannot be avoided. From birth, our existence as a human body is shaped by our connections to other bodies (human or otherwise) that, like us, will die. From birth we are at once already survivors and preparing ourselves to be survived (Derrida 1996; Brault and Naas 2001). Mourning, and the resulting work, then, is an unyielding and ever-present condition of life, the labours of which we partake in ceaselessly, interminably, and inconsolably (Derrida 1996). The work of mourning begins before death, with the knowledge that "we" will be surviving "others" (whomever and whatever those "others" are). In this way, mourning is also the opportunity to continually engage with death, with loss, and with those who have come before while we are still alive (Brault and Naas 2001; Kirkby 2006; Engle 2007).

If mourning is a work for the living and from which we cannot escape, is it ever fully finished? While there are different answers to that question, Freud laid the foundations for a belief that healthy mourning would eventually come to completion, and this theory still characterizes much discourse on mourning. Freud offered a psychoanalytic framework of mourning that reported successful mourning was about being able to substitute one object for another (this changed in his later works to the full incorporation of loss) (Butler 2004). Freud's theory of mourning was premised on a dichotomy between mourning (a process of grief over loss of a love object that had the potential to become resolved) and melancholia (mourning that was not "successful" and had become pathological). For Freud, mourning is a "normal" and "regular" response to the reaction of losing a loved one or a loved object, which sparks a process of painful experiences and withdrawal from one's daily activities. Conversely, melancholia can linger on in an unhealthy or narcissistic way, holding on to the pain and suffering (Kristeva 1989; Freud 2007). Successful mourning, then, emerged if the mourner no longer felt pain, no longer withdrew from regular activities, and found something new to replace what was lost – and avoided the temptations of melancholia.

Mourning, "profound mourning, the reaction to the loss of someone who is loved, contains the same painful frame of mind, the same loss of interest in the outside world – in so far as it does not recall him – the same loss of capacity to adopt any new object of love (which would mean replacing him) and the same turning away from any activity that is not connected with thoughts of him" (Freud 2007, 20). As Catriona Sandilands explains in chapter 6 (145), in Freud's view, "loss involves the rupture of a libidinal attachment; the beloved object is no longer available for, or rejects, our love. As a way of preserving that object, we incorporate it into ourselves, internalizing the attachment: the ego holds on to the object by devouring it and making it part of itself, substituting narcissistic energy for cathectic, an inner attachment for a new, outer one."

Jacques Derrida builds on and moves beyond Freud, articulating an understanding of mourning that does not strive to return to a pre-loss state. As Helen Whale and Franklin Ginn describe in chapter 4 (98): "Mourning in this Derridean sense is not the traditional psychoanalytic return to 'normal' after a loss, but a diminishment of the prospects for becoming. Derrida suggested that to internalize someone in memory denies them their independence, it erases their autonomy, but on the other hand not to remember them means that we lose them completely. In the traditional psychoanalytic understanding of mourning, Derrida argues, we end up losing the deceased a second time, as they are acknowledged, but then put away again 'in us' – their otherness is removed ... The solution, Derrida suggests, is not to transcend or reconcile this tension, but to recognise it, and to retain a sense of the 'irretrievability' of the lives of others." From a Derridean perspective, therefore, "successful" mourning is not about internalizing or replacing the other; rather, mourning is about recognizing that all we have to give to mourning, to the dead, to what we have lost, is in our own living and our own acts of mourning and remembering (Brault and Naas 2001). For Derrida (1996), this work is about responding to the responsibility posed by the deceased and through grief, and of responding through our ethical and political choices. Mourning is both a necessity of life and a call to responsibility to engage with the dead and with what was lost (Derrida 1996; Kirkby 2006), and it carries with it a requirement of response through our own lives and actions (Naas 2003).

Mourning, then, is work to which we must always attend and which we must always share with others, and it does not finish while our own body is alive. In mourning, we not only lose something that was loved, but we also lose our former selves, the way we used to be before the loss. We are changed internally and externally by the loss in ways that we cannot predict or control, and in ways that may be disorienting, surprising, or completely unexpected. Through this mourning-as-transformation we are open to, continually exposed and vulnerable to, these bodies through the potential for loss and our subsequent grieving. In this understanding of mourning, we are also continually seized by unexpected responses to loss for which we can little prepare, and which continually compound through subsequent experiences with loss and grief. These responses to loss can leave us changed in ways we could not have imagined, and hold the possibility of leaving us more open to other bodies, to grief, and to our transcorporeal connections with all bodies.

Or, conversely, we can become closed off, desensitized to the suffering and destruction of other bodies – a desensitized condition that the work of mourning can challenge and disrupt.

## Mourning as Politically and Ethically Transformative

Mourning is a cultural, political, and ethical practice. It is something shared across cultures, and in response to grief-producing events, such as the death of a loved one or personal tragedy, or the loss and devastation from large-scale events to which we may or may not be personally connected: wars, genocide, terrorist attacks, epidemics, famine, natural disasters, extinction, and the loss of biodiversity. This grief is experienced individually and privately through one's own emotional and corporeal responses, but is also often shared publicly with others, through collective expressions or gatherings of mourning (such as funerals, memorials, protests, eulogies, obituaries, and vigils). Grief can be shared by a relatively small group of familiar people in memory of an individual, or on a large scale in response to shared grief by a collective group of strangers in a public setting through public outpourings (for example Columbine, 9/11, Matthew Shepard, and the Montreal Massacre at the École Polytechnique) and through mourning at public memorials (for example the AIDS Quilt, Grave of the Unknown Soldier, Korean War and Vietnam War memorials, Ground Zero, Highway of Heroes, Hiroshima Peace Memorial, the Killing Fields Museum).

While Freud and Derrida's thinkings are woven throughout this collection, it is clear that many of our authors took inspiration from Butler's work on mourning, particularly in *Precarious Life: The Powers of Mourning and Violence* (2004). Butler builds on, and then moves away from Freud and Derrida to postulate a theory of mourning premised on shared vulnerability, transformation, and the questioning of what bodies count as mournable, and why. Butler's work exposes the underlying connections among mourning, grief, and larger geo-political, socio-economic, race, and class issues to expose the fact that who and what we do not mourn tells us as much about ourselves as who and what we do mourn. There are bodies that simply do not matter, or have been, as Ashlee Cunsolo explains in chapter 7, "disproportionately derealized from ethical and political consideration in global discourse: women, racial minorities, sexual minorities, peoples of different religions, certain ethnic groups, economically and politically marginalized groups, and Indigenous peoples, to name a few" (170).

By envisioning an *end* to mourning, a Freudian perspective misses the many potent political and ethical acts that can emerge from mourning; Butler's work helps us move beyond Freud's framework, for mourning, of full interiorization and replacement of the love object to a more politically and ethically salient perspective of mourning and its associated work. For Butler (2004, 21), "successful" mourning and grieving does not come from the full substitutability or the forgetting of what was lost; rather, mourning is about transformation: "one mourns when one accepts that by the loss one undergoes one will be changed, possibly forever. Perhaps mourning has to do with agreeing to undergo a transformation (perhaps one should say submitting to transformation) the full result of which one cannot know in advance. There is losing, as we know, but there is also the transformative effect of loss." Beyond being a necessary condition of life and relations, both Derrida and Butler have argued that mourning is also a potent political force. For Derrida, "there is no politics without an organization of the time and space of mourning" (Derrida 1993, 61), without the recognition of our ethical and political responsibilities to the other through the recognition of the fragility and vulnerability of our own life. For Butler (2004, 22), the process of mourning "furnishes a sense of political community of a complex order, and it does this first of all by bringing to the fore the relational ties that have implications for theorizing fundamental dependency and ethical responsibility." This "we-creating" capacity of mourning (Butler 2004) is what exposes our relations to and connections with others – whether we know those others or not – and where the potential for enhancing individual and collective resilience to loss through a shared capacity to grieve, to suffer, and to mourn. Mourning can unite, and grief over a shared loss or something integral to one's self can be a powerful political motivator and unifier. Grief is also unique in its capacity to reach across cultures, languages, and differences and connect with others through recognition of the shared pain and suffering over the loss.

Regardless of how we experience loss and respond with and to mourning, we are characterized in part by the continual loss of lives and bodies around us through events both in our control and beyond our control. We are also characterized in part by who and what we grieve and, just as importantly, by who and what we do not grieve. As Butler (2004, 46) explained: "If I understand myself on the model of the human, and if the kinds of public grieving that are available to me make clear the norms by which the 'human' is constituted for me, then it would seem that I am as much constituted by those I do grieve for as by those whose deaths I disavow, whose nameless and faceless deaths form the melancholic background for my social world."

There are, in many cases, deaths and loss that go unnoticed or unmourned by ourselves and others, or who do not seem to matter in discourses of mourning – human and non-human bodies. And yet while we may not explicitly mourn, we are still shaped by those myriad losses and are affected consciously or unconsciously. From this perspective, mourning not only exposes our connections to others – human, animal, vegetal, mineral, or hydrological – and provides an opportunity to connect to ourselves and others through loss and shared vulnerability, it also provides political and ethical opportunities to expand discursive spaces to include bodies not mourned in dominant discourse, and to encourage individual and collective action. We are all implicated in this process of "being-towards-death," and this frames not only the boundaries and outlines of our lives, but also makes possible our interactions in the world and our interactions with other bodies, whomever and whatever those bodies may be (Derrida 1996). Neither "being-towards-death" nor mourning can ever be avoided.

Herein lies the work of mourning: we are born into this work, as we are born already preparing to say goodbye; we are already survivors, and we are already preparing ourselves to be survived (Derrida 1996). Mourning is simultaneously individualizing and unifying; it can be experienced quietly and privately through one's own emotional and corporeal responses, making one feel separate from others, or it can be experienced collectively, uniting us with bodies, places, and events beyond ourselves, and even beyond our immediate geographic locale or personal spheres. These tasks and processes of mourning take over the lives of mourners for varying temporal lengths, to varying degrees, and interrupts normal activities and habits, while attempting to rebuild a new self in the wake of the loss (Dubose 1997). It is precisely because mourning has this dualism that it can also hold the potential for political and ethical action. We can at once be moved and changed by our own grief and the grief of others – whether we have directly experienced an event or not. In this grief, we can come together through the shared experience of understanding and feeling loss, and allow for a shared transformation.

Mourning "is always a political act. Although it is frequently viewed as a private experience, indeed, an experience that flirts with solipsism, mourning is all about ethical, political, and ontological connections ... Mourning is a way of making connections, of establishing kinship, and of recognizing the vulnerability and finitude of the other" (Stanescu 2012, 568, cited in chapter 8, 198). Mourning, then, can take on an overt political and ethical form, becoming what Clifton Spargo (2004) identified as resistant mourning. As both Spargo and Patricia Rae (2007) explain, resistant mourning is mourning that refuses consolation, and consciously chooses to hold on to the feelings of pain and grief to spur a sense of responsibility for the loss. For Spargo, this resistant mourning becomes an ethical protest against the larger structures of injustice and oppression that trivialize and minimize the death and loss of some bodies.

A strong example of resistant mourning can be witnessed through the response to the AIDS epidemic, where families and activists mourned publicly to mobilize their grief into something that worked to redefine the AIDS body as grievable (Rae 2007; see also Cunsolo's chapter 7, 176). For Rae (2007), resistant forms of mourning, at their core, leave "mourning unresolved without endorsing evasion or repression; indeed, they portray the failure to confront or know exactly what has been lost as damaging. They encourage remembering where memory has been repressed, and they expose the social determinants for troublesome amnesia. At the same time, they resist the narratives and tropes that would bring grief through to catharsis, thus provoking questions about what caused the loss, or about the work that must be done before it is rightly overcome. They raise questions about the social forces that have prevented the work of mourning from being accomplished, and they offer alternatives to the consolatory strategies that have been widely deployed and that threaten to introduce a whole new round of loss and grieving"

(2007, 22–3). Within this understanding, thinking with grief has the capacity to act as a form of resistance to the pressures, forces, and processes that are underpinning current practices of environmental destruction and commodification. Bodies – our own and others – are increasingly the subject of politicization and materialization, and mourning can stand as an opposition to both death and the commodification of the living. This "resistant mourning" (Rae 2007; Spargo 2004) or "activist melancholia" (Rae 2007) can encourage positive social change as we accept a sense of responsibility for both past losses and losses to come. In resistant mourning, then, experiences of loss and associated grief are mobilized to expose systematic marginalization, political injustices, and systemic violence.

This grief and this mourning are not combined with anger but are, rather, mobilized and supported by unity through the recognition of our shared vulnerability (Butler 2004) and through the hope and collective strength within the context of devastating loss bringing, as Lisa Kretz indicates in chapter 10, a hopeful lens to the process. For, in this work, there is the possibility that positive emotions and radical hope can be supplements to the work of mourning, a counterpoint to "the dry grief of an endless political rage" (ibid., xix) by "furnishing a sense of political community of a complex order" premised on "bringing to the fore the relational ties that have implications for theorizing fundamental dependency and ethical responsibility" (ibid., 22).

## The Collection: Moving Mourning beyond the Human

This collection presents compelling and important works that help us think with, through, and about mourning that moves beyond the human. It contains heartfelt and emotional stories and narratives that testify to the visceral, sensorial, and sensuous connections we all share with more-than-human bodies. In light of the political and ethical capacities of mourning and the associated work, this collection takes up the challenge to extend the theoretical and lived experiences of the work of mourning to non-human bodies, singular or collective, be they animal, vegetal, mineral, hydrological, or geological, and to consider new terrain in environmental thought and ethics and new points of political, ethical, and social action.

The works enclosed seek to disrupt the dominance of human bodies as the only mournable subjects, and to argue against "the human that falsely occupies the space of the universal and that functions to exclude what is considered nonhuman (which, of course, includes the immense majority of human beings themselves, along with all else deemed to be nonhuman) from ethical and political consideration" (Calarco 2008, 13). The authors build on developments in multispecies theories and post-humanist ethics by scholars such as Donna Haraway (2008), Timothy Morton (2010), Deborah Bird Rose (Rose 2008, 2011a, b), and Thom van Dooren (2011, 2014), and do so through the lens of bringing grief work to the more-than-humans with whom we share the planet, and who so often are derealized from the realm of grievable. This collection, then, tries to make some space where the authors and the readers can join together to think with and through moving mourning beyond merely the human.

Yet this collection also does this work with caution. While we may already be experiencing mourning related to non-humans, and while we may want to extend this mourning beyond the human, Smith (2013) reminds us that although it is hard to imagine the loss of a human, it is harder still to think through the loss of non-humans, particularly if the non-humans we are mourning are large assemblages or systems, such as a body of water or an ecosystem or a forest. As Nancy Menning writes: "We don't grieve abstractly; we mourn particular losses of people, places, animals, objects, and ideas to whom and to which we are attached ... Ecological losses differ in important ways from human deaths. In particular, we are often complicit in these losses, if only by virtue of living in the Anthropocene. We must mourn not only what we have lost, but also what we have killed or otherwise destroyed ... When one feels complicit (directly or indirectly) in the loss being mourned, guilt entwines with sorrow, complicating the grieving process" (chapter 2, 39). This ecological work of mourning will, necessarily, look different. It will feel different. And it will require from us different forms of commitment, different attention, and different ways of thinking. It will require us to engage with ourselves, with others, with non-humans, and with emotions differently. And if we take up this challenge, it may compel the creation of new understandings of ethico-political strategies for mourning beyond the human based on how non-human bodies are created, valued, understood, populated, realized, and derealized.

## The Chapters Within

In efforts to understand impacts on more-than-human bodies, traditional scientific forms of habitat assessment have relied on counting species and individuals within each species. In the early twentieth century, the use of acoustic devices allowed scientists to single out the song of individual bird species. However with this method the single voice of a bird is decontextualized from its mileu. Instead, Bernie Krause's assessment of diversity is based on *soundscape*, a concept first developed by Canadian composer and naturalist R. Murray Schafer in 1977. Soundscape, drawing on what we hear and sense in the acoustic landscape, allows for an immersion into habitat that counting can never capture. Krause has shown that when habitat is under stress due to human disturbance, the acoustic structure of that habitat breaks down. He demonstrates how both humans and non-humans react to the breakdown of soundscape structure – "lush biophonies" – and the associated sense of the loss of place. This deficit, Krause tells us, affects well-being for all living organisms and can trigger feelings of loss and grief. Krause takes the reader on a visceral and affective acoustic journey – demonstrating the deep relationships among humans, soundscapes, and ecological bodies – and illustrates the intense grief that can be experienced and witnessed when there is a loss in the acoustic realm, signalling a loss of more-than-human bodies and worlds.

Through a number of cultural examples from West Africa, traditional Judaism, Tibetan Buddhism, and Shiite Muslim rites, Nancy Menning shows us how mourning rituals contribute to healing. From these examples, Menning suggests ways by which we can conceive of culturally appropriate possibilities for extending such rituals to also mourn ecological losses. She argues that rituals ease grief, resolve guilt, and can prepare us for the work of environmental activism. Mourning rituals offer structure and the means for interaction when we are bewildered and disoriented in our grief. Mourning not only eases our grief, she states, it can also direct us to creating a better future.

In his chapter on the kinship relations of the Lakota with the buffalo, Sebastian Braun takes us to the Great Plains of the United States. The loss and absence of buffalo in this landscape is symbolic of a historical trauma for the Lakota, and for the landscape. Braun argues that people feel related to their environment through kinship relations, and that relationships with non-human entities form the basis for any relationship with one's environment. Kinship implies obligation, and obligation implies responsibility. For the Lakota, the killing of a buffalo requires an obligation to mourn that animal, ensuring an ongoing fundamental relationship with buffalo as a species. "The incapacity to mourn," Braun (chapter 3, 86) writes, "is the indicator of not simply a loss of empathy for our environment, but the loss of kinship relations, of social meaning in general."

Helen Whale and Franklin Ginn have also researched the implications of the loss of a particular species – or more accurately the absence of that species – in their work on the decline of house sparrows in London, UK. When a bird species disappears from its former place, how do we experience that absence? With the London House Sparrows Parks Project as a context for interviewing local residents, Whale and Ginn explore how the decline of the house sparrow is experienced through mourning, landscape, and the revelation of absence. This absence is not just about loss but also about the new place that is created as a result – a haunted and spectral sparrow place. Whale and Ginn show that signalling a loss or absence can also be generative of place – both a place that *is* and a place that *once was*. By making the invisible visible, the House Sparrows Parks Project reifies the human connection with the more-than-human world(s) and conjures not only mourning for a lost species but also for the relational places that are associated with house sparrows – i.e, meadow, hedge, flock, childhood, and London itself.

Discussion of the shared, and now nostalgic, experience of sweet-scented boronia in South-West Australia, as a herald of spring, leads John Charles Ryan to the concept of floratopaethesia – meaning the sense of place fostered through the sensory experience of flora. South-West Australia's biodiversity is internationally renowned as one of the world's most botanically diverse Mediterranean ecosystems. The loss of boronia in the landscape – and the accompanying loss of the heady perfume of their spring flowers that was a tradition in Perth – triggers memories for Ryan's interviewees of a long-established ecological intimacy that has been lost due to anthropogenic landscape change. Ryan writes that his interviewees reveal an impoverishment of their sensory lifeworlds, memory networks, and emotional well-being due to a severe reduction of plant species diversity. The result is an ecology of mourning that is complex, inclusive, and evocative of place. The mourning of a plant species is not just about the loss of that particular species, therefore, but also about the loss of the milieu in which it had been ecologically entangled, an entanglement in which Ryan's interviewees had also been immersed through their sensory appreciation of boronia.

Catriona Sandilands's chapter focuses, in part, on her own well-being by sharing with us the wrenching loss of both her marriage and her mother in a short time period, and the resulting dislocation of her place in a landscape. She tells us of the trauma, loss, and anger she felt with all that had been lost. For Sandilands, the loss of these attachments contributed to her experience with depression, and the resulting links that emerged to ecological loss. "I had lost my place, and part of my self along with it," she writes (chapter 6, 150), and links her personal experience with depression to a larger societal malaise. In discussing these linkages she leads us through the writings of John Bentley Mays, Sylvia Plath, and Anne Michaels to reveal our interconnectedness with the more-than-human and particularly through the loss of landscape – we are connected by these mutual losses.

Ashlee Cunsolo argues that mourning losses due to climate change can be a unifying work, through a collective experience of shared grief. If we move from grief to the individualized work of mourning and then to shared mourning, the resulting transformative experience will leave us changed in ways that are unanticipated, but also hopeful. Cunsolo's discussion of public mourning for the loss of the more-than-human points to the potential for productive political mobilization in the face of landscape changes induced by climate change. She recognizes the resulting impacts on us and on landscapes, and offers unification through shared and mobilizing mourning.

Jessica Marion Barr presents her research into the creative practice of the modern elegists of the twentieth century, who created resistant art in the face of massive destruction during two world wars. As an artist, Barr's work engages in ecological elegy as a deliberate form of mourning that is a hopeful attempt to shape a better ecological future. She also presents the work of other ecologic elegists in her chapter on the art and ethics of ecological grieving. In response to and in resonance with resistant mourning, Jessica Marion Barr takes up the concept of "proleptic elegy": elegies that are produced by artists and poets in response to coming events or the possibility of more tragedy and violence, including, for Barr, environmental destruction and extinction. Taking up the understanding that proleptic elegies "could justly be added to the arsenal of resistant modes of mourning compiled in recent years by activists looking for social hope in devastating loss" (Rae 2007, 229), Barr moves beyond human bodies to examine the anticipatory grief and resistant forms of mourning that proleptic *ecological* elegies entail, "grieving and warning about the kinds of future losses that could occur" (chapter 8, 197), and bringing important attention and awareness about ecological loss and destruction through visual, audio, and literary narratives full of pain, affect, emotions. Barr argues compellingly that this art is hopeful and has the power to cause us to reconsider our relationships with ecological others to extend grievability to those others and then galvanize us to take positive action.

Andrew Mark and Amanda Di Battista's chapter brings together disparate ideas about environmental loss; in doing so, they consider the role of a podcast on ecological loss and mourning as a sounded ecology. They argue that the podcast, as an acoustic form, creates a useful way to explore environmental loss and mourning, and that the phenomenological experience of the sounded information changes how we know and experience *place*. All the participants who were interviewed for the podcast called for a politicization of mourning – whether mourning for humans or for the more-than-human – by focusing on particular losses through ritual, creative, and artistic practices. Mark and Di Battista argue that the podcast form allows for fresh perspectives on environmental mourning, and it is also a creative way to confront environmental problems.

In assessing how emotion can empower such positive action, Lisa Kretz examines our capacity for appropriate emotional and moral responses to ecological mourning. She states that the possibilities for the emotional shaping of our lives and the potential for intentional emotional self-configuration to deal with environmental loss are under-utilized, and are potentially extremely powerful sources for motivating positive environmental action. Conversely, the dismissal of these emotions serves as an effective political tool to deny oppressed groups, whereby marginalized viewpoints are considered to be irrelevant. This dismissal outlaws such emotions so that those who hold these emotions are disavowed, disempowered and silenced. In her argument, Kretz leads us through justified anger to justified hope; in generating justified hope, Kretz infers that it is possible to come together to create community capacity in responding to ecological harm. Through hope, Kretz argues we can create radical and positive change.

In the creation of a psychoterratic typology that captures earth-related emotions and feelings as they have evolved in nature and place writing, Glenn Albrecht revisits his concept of solastalgia, which is the "homesickness you have when you are still at home" – a particular form of distress related to negative changes in one's home landscape. Moving beyond solastalgia, Albrecht argues that we need a whole new taxonomy of language to understand and respond to rapid environmental degradation and loss. This renaming is important, for, as he argues, "the explicit giving of names to that which had previously been intangible or subliminal is empowering and enables those who participate in the named drama to be collaboratively creative and engage in a community of scholarship, politics and criticism" (chapter 11, 302). Albrecht argues for an expansion of psychoterratic typology beyond established terms like ecoanxiety to start to include terms such as "memerosity" (pre-solastalgic state of worrying about the degradation of place), and to link these new terms to an understanding of place-based grief and mourning. Finally, he argues for the term "symbiocene" to replace our overused (and therefore often devoid of meaning) "environment," suggesting that it will also allow us to more clearly articulate the interrelationships and interdependencies of all bodies, human or more-than-human.

The collection concludes with a reprint of the convocation address that Canadian poet Patrick Lane gave to University of Victoria students in 2013. In his address, he tells a story of a particular time and place, and of a particular response to nature that was, for him, a searing and disorienting experience. He speaks of the dark forces of destruction that surround us – that this, too, is the world we have inherited. He asks, not without hope, "how do you want this story to end?"

## A Renewed Form of Ecologically Based Mourning

The works contained within this collection illustrate that no death is irrelevant, whether it is the death of a human body like our own or a body beyond the human. As is made clear in this collection, we need a new form of mourning, a new form of ecological ethics and politics to mourn beyond our species, beyond human bodies, to expand the boundaries of what constitutes a mournable body and what does not. This is not only a fecund area for future work but also an urgent point of action as we confront the increasing causalities to the seemingly relentless march of environmental degradation and destruction. As Burton-Christie (2011, 30) writes: "the ability to mourn for the loss of other species is, in this sense, an expression of our sense of participation in and responsibility for the whole fabric of life of which we are a part. Understood in this way, grief and mourning can be seen not simply as an expression of private and personal loss, but as part of a restorative spiritual practice that can rekindle an awareness of the bonds that connect all life-forms to one another and to the larger ecological whole" (cited in chapter 3, 74).

Writing about, thinking with, and living through grief has its challenges and its pain; it has the potential to exact a toll if we cannot find outlets or ways to share. This is, in many ways, why we wanted to create this collection, to give voice to what many of us are already feeling, already struggling with, and to open up the conversations on these topics so we can start considering what it means to be living in a time where we are experiencing unprecedented loss, death, and destruction – a time where we are facing our shared vulnerability in unexpected and deeply troubling ways while simultaneously having the capacity for hope, joy, and individual and collective action. There are already more-than-human bodies that have passed, and remain only in memory or in memorial, and there are those that are passing right now, at this moment, and cannot be saved. And then there are those that are still here, with us, in various forms of being-towards-death, and to those bodies we have a responsibility. We owe them our grief. And we owe them our action. It is here that we take our start from Lisa Kretz when she writes: "mourn then, what cannot be saved, but defend what remains and can be" (chapter 10, 275).

While the works in this collection are diverse and represent a broad spectrum of approaches, content, and ideas – from mourning landscapes and soundscapes to resistant mourning and elegiac grieving to mourning sparrows, *Boronia*, and gardens – this is but a start in what we believe is an important conversation within environmental thought, theory, and action, and what we hope will be a growing field of analysis and inquiry. We envision this collection as a platform for further conversations and dialogues on the complex interplay between "environment" and "mourning," and the multitude of voices, topics, areas of thought, points of action, and epistemological and ontological shifts that help us think about what this loss tells us about ourselves and others, human and more-than-human. Yet there are still noticeable absences in this collection; we need to think about how environmentally based mourning intersects with other forms of mourning, and how issues of race, class, socio-economic status, geographic positionality, geo-politics, colonial legacies, and transnational corporatization (to name a few) interact with and affect how we understand not only the environment, place, and more-than-human bodies, but also how we understand and undertake the associated mourning. We need to continue to consider more deeply how mourning beyond-human-bodies can become a resource for political and ethical change, and consider how we can leverage this ecologically based grief in productive, meaningful ways. We need to consider the types of mourning and grief that emerge from extinction, from loss of biodiversity, from natural disasters, and from displacement. And we need to consider how the work of mourning may expand or inform the work being done in post-humanist ecological ethics and politics.

We hope that you find this volume as stimulating, moving, inspiring, and provocative to read as it was for us to prepare, and we hope that it provides both intellectual and emotional foundations upon which you can start building, expressing, and sharing your own environmental grief narratives, your own theoretical frameworks for understanding environmental change and degradation as the work and labours of mourning, your own political and ethical frameworks for productive and positive action through individual and collective expressions of ecological grief.

The stories that we tell about loss and mourning are important; if we can also connect our grief and pain to our hope, emotions, and actions, we may well be able to move beyond being stuck in mourning or lost in "endless political rage," to move into new domains and terrains of interspecies interaction and connectivity. The authors in this collection remind us that, in the work of ecological mourning, theories will only get us so far – emotions may just carry us the rest of the way.

There is much work to be done. We hope you will join us and share your voices and grief experiences beyond the human, bridging species and bodies in communities premised on ecological mourning and ethico-political mobilization. For, as Derrida (1989, xvi) wrote about mourning, "speaking is impossible, but so too would be silence or absence or a refusal to share one's sadness."

*Sydney, Nova Scotia, and Guelph, Ontario*

## NOTES

1 Much of the next two sections was originally published in Ashlee Cunsolo Willox's (2012) article, "Climate Change as the Work of Mourning," published in *Environmental Ethics*, and appearing in this collection as chapter 7. That article was one of the inspirations for this collection, and laid some of the intellectual foundation for our work; as such, to avoid repetition, and in order to create a cohesive collection, we have removed some of the more introductory pieces from chapter 7, and inserted them into the introduction in revised form.

## REFERENCES

Brault, Pascale-Anne, and Michael Naas. 2001. "To Reckon with the Dead: Jacques Derrida's Politics of Mourning." In *The Work of Mourning*, edited by Pascale-Anne Brault and Michael Naas, 1–30. Chicago: University of Chicago Press.

Burton-Christie, Douglas. 2011. "The Gift of Tears: Loss, Mourning and the Work of Ecological Restoration." *Worldviews* 15: 29–46.

Butler, Judith. 2004. *Precarious Life: The Powers of Mourning and Violence*. New York: Verso.
Calarco, Matthew. 2008. *Zoographies: The Question of the Animal from Heidegger to Derrida*. New York: Columbia University Press.
Derrida, Jacques. 1993. *Aporias*. Translated by Thomas Dutoit. Stanford: Stanford University Press.
– 1996. "By Force of Mourning." *Critical Inquiry* 22 (Winter): 171–92.
– 1989. *Memoires for Paul de Man*. Translated by Cecile Lindsay, Jonathan Culler, and Eduardo Cadava. New York: Columbia University Press.
Dubose, J. Todd. 1997. "The Phenomenology of Bereavement, Grief, and Mourning." *Journal of Religion and Health* 36, no. 4: 367–74.
Engle, Karen. 2007. "Putting Mourning to Work: Making Sense of 9/11." *Theory, Culture, and Society* 24, no. 1: 61–88.
Freud, Sigmund. 2007. "Mourning and Melancholia." In *On Freud's "Mourning and Melancholia,"* edited by Leticia Glocer Fiorini, Thierry Bokanowski, and Sergio Lewkowicz, 19–34. London: International Psychoanalytical Association.
Haraway, Donna. 2008. *When Species Meet*. Minneapolis: University of Minnesota Press.
Kirkby, Joan. 2006. "'Remembrance of the Future': Derrida on Mourning." *Social Semiotics* 16, no. 3: 461–72.
Kristeva, Julia. 1989. *Black Sun: Depression and Melancholia*. Translated by Leon S. Roudiez. New York: Columbia University Press.
Morton, Tim. 2010. "The Dark Ecology of Elegy." In *The Oxford Handbook of the Elegy*, edited by Karen Weisman, 251–71. Oxford: Oxford University Press.
Naas, Michael. 2003. "History's Remains: Of Memory, Mourning, and the Event." *Research in Phenomenology* 33: 75–96.
Rae, Patricia. 2007. "Double Sorrow: Proleptic Elegy and the End of Arcadianism in 1930s Britain." In *Modernism and Mourning*, edited by Patricia Rae, 213–38. Lewisburg: Bucknell University Press.
Rose, Deborah Bird. 2008. "Judas Work: Four Modes of Sorrow." *Environmental Philosophy* 5, no. 2: 51–66.
– 2011a. "Flying Foxes: Kin, Keystone, Kontaminant." *Australian Humanities Review* 50: 119–36. www.australianhumanitiesreview.org/archive/Issue-May-2011/home.html. Accessed 30 April 2013.
– 2011b. *Wild Dog Dreaming: Love and Extinction*. Charlottesville: University of Virginia Press.
Smith, Mick. 2005. "Citizens, Denizens and the Res Publica: Environmental Ethics, Structures of Feeling and Political Expression." *Environmental Values* 14, no. 2: 145–62.

– 2013. "Ecological Community, the Sense of the World, and Senseless Extinction." *Environmental Humanities* 2: 21–41.
Spargo, R. Clifton. 2004. *The Ethics of Mourning.* Baltimore: The Johns Hopkins University Press.
Stanescu, James. 2012. "Species Trouble: Judith Butler, Mourning, and the Precarious Lives of Animals." *Hypatia* 27, no. 3: 567–82.
Van Dooren, Thom. 2010. "Pain of Extinction: The Death of a Vulture." *Cultural Studies Review* 16, no. 2: 271–89. http://epress.lib.uts.edu.au/journals/index.php/csrj/article/view/1702. Accessed 30 April 2013.
– 2014. *Flight Ways: Life and Loss at the Edge of Extinction*. New York: Columbia University Press.

# 1 Mourning the Loss of Wild Soundscapes

## *A Rationale for Context When Experiencing Natural Sound*

BERNIE KRAUSE

### Introduction[1]

Historically, natural soundscapes have been recorded in fragmented, decontextualized formats, the emphasis focused on the abstraction of single species' animal voices intentionally removed from their more holistic and informative acoustic fabric. This older model, first expressed in the late nineteenth century when recording technologies were in their infancy, became an act of faith by the 1930s when ornithologists first applied the parabolic dish primarily to the capture of bird song and calls. The 1935 recording of the ivory-billed woodpecker in a Georgia swamp by Arthur Allen and Peter Kellogg from Cornell's Macaulay Library of Natural Sound fixed a course that has been followed rigidly ever since. When hearing these recordings, it's a bit like trying to understand the marvels of Beethoven's Fifth Symphony by listening to a solo violin part abstracted from the context of the symphonic performance. Hearing the world in this decontextualized way distorts our sense of perspective and, as a result, introduces degrees of stress of which we are often unconscious; instinctively, many of us sense something missing. The single-minded focus on this procedure has obviated the perception of a much more inclusive and resonant comprehension of sounds emanating from the natural world, one that – although very late in the process when so many natural soundscapes have disappeared entirely – still conveys large amounts of significant content through its more inclusive signal.

The narrower and fragmented scope of early collections is, however, beginning to change as the information revealed in the larger contextual soundscape recordings compels us to examine the larger frame of reference more closely. Among the most persuasive observations are those occurrences of depression, sadness, and changes in behaviour that occur in humans and non-humans alike, primarily as a result of loss of or radical shifts in the density and diversity expressed through the vocal organisms that inhabit more stable landscapes. But in order to comprehend the synapse between natural sound as it occurs in its wild states, and animal interaction and response, it is necessary to first grasp how the concept of soundscapes work.

## Current Knowledge about Soundscapes

The soundscape model was first defined – as "all of the sounds that reach the human ear from whatever source" – by R. Murray Schafer, Canadian composer and naturalist, in his seminal 1977 publication, *Tuning of the World*. Around the turn of the millennium several of us working in the nascent field of soundscape ecology realized the need to expand on that definition as well as the wider bioacoustic lexicon. This meant expanding the scope of observation to include all the sources of sound that occur within the soundscape. The entire soundscape model features three components: *biophony*, *geophony*, and *anthropophony*. Typically, *biophonies*, the most important of these collective signals, have evolved from relatively undisturbed habitats and are comprised of various combinations of non-human, non-domestic insects, reptiles, amphibians, birds, and mammals depending on time of day, night, season, or weather dynamics. The *geophony* is non-biological natural sound coming from the effects of wind, water, and earth movement. And finally, *anthropophony* is human-generated sound – all of the signals that humans introduce into our environments – whether controlled, as in music, or entropic, as in the chaotic noise we create with our electro-mechanical implements.

Traditional forms of habitat assessment have usually been approached by visually counting the numbers of species and the numbers of individuals within each species across the range of a given biome.

However, by comparing data that ties together density and diversity of what we hear, we are now able to illustrate habitat fitness outcomes that meet or exceed those of other methods. Visual capture explicitly frames a limited frontal perspective while soundscapes widen that scope to include a full 360 degrees, completely enveloping us with useful information. Geophonies and biophonies often work together, communally expressed as inherent acoustic sound marks of a particular habitat signalling intrinsic levels of health and indicating whether its collective voice exists within an expected range of dynamic equilibrium (i.e. a working definition of a thriving habitat). When those conditions occur and connect, the voices of the natural world are expressed through distinctive patterns of cohesion and organization – acoustic unities where each type of organism tends to establish bandwidth within the frequency and/or temporal spectrum so that its voice can be heard unimpeded and unmasked by others – conditions that are illustrated graphically in spectrograms, or images of sound that show frequency and amplitude over time.

When a habitat is under stress because of human intervention, like resource extraction, landscape alteration, global warming, pollution, anthropophony, or the extreme effects of a particular weather or geologic event, this acoustic structure breaks down and what appears on the page is an illustration of entropy, a form of chaos showing little or no bioacoustic discrimination between types of signal-producing organisms. Many life forms, especially those whose existence and behaviour depend on flourishing within the frame of healthy wild habitats, will tend to react to the breakdown of soundscape structure in ways that although often overlooked are nevertheless notable. Examples from both the human and non-human worlds follow.

## Psychological Effects Caused by the Loss of Wild Soundscapes

Over the course of many years, I have witnessed the intense psychological responses that humans have had when their soundscapes were altered, disrupted, or lost. Often similar to a sense of homesickness or loss because a familiar place or soundscape was rendered unfamiliar –

resonating with Glenn Albrecht's concept of solastalgia in chapter 11 – people respond with myriad emotions, including grief, mourning, sadness, depression, and desolation.

### The Wy-am People and Celilo Falls

Years ago, while I was working in the Northwest along the Columbia River, a local resident told me the story of a nearby Native American group whose communal life had completely revolved around a waterfall – a sound from their creation story. The waterfall animated their lives as a group and sustained them through every generation. When I told the local that I was curious to learn more, he introduced me to tribal member Elizabeth Woody.

According to Woody, Elders in the Wy-am tribe tell of a period spanning thousands of years when they fished all year long at Celilo Falls, just west of the Columbia River's midway point (*Wy-am* means "echo of falling water"). So central were the falls to the tribe that the Celilo was considered a sacred voice through which divine messages were conveyed. Each season the wide, vital river at the cascade provided lots of fish – in the spring, chinook salmon; in the summer, chinook and bluebacks; in the fall, chinook, steelhead, and coho. When the catch was good, tribal members could harvest a tonne of fish a day. For little more than the cost of a couple of balls of twine, tribesmen could quickly supply both close and extended family with fish for a year.

On the morning of 10 March 1957, the US Army Corps of Engineers, hoping to improve navigation on the river, ordered the massive steel gates of the newly built Dalles Dam shut tight, strangling the natural downstream flow of the river. Six hours later, the sacred waterfall and fishing site of the Wy-am, eight miles upstream, were completely submerged. Although they had been forewarned, the Wy-am elders stood on the riverbank, astonished, watching as a way of life that had flourished for centuries disappeared in less than a day. There wasn't a dry eye on the banks near Celilo, the small namesake village on the river's edge. And yet the Elders were not weeping just for the loss of salmon. They wept because the river no longer lent its wise voice to the community. The submerged Celilo Falls were dead silent. "It was a place revered as one's own mother," said Elizabeth Woody. "I [now] live with the …

absence and silence of Celilo Falls much as an orphan lives hearing of the kindness and greatness of her mother" (Krause 2012).

Many Native American groups have central to their mythology stories about the sounds of creatures that inhabited their landscapes. In 1971, I was on assignment to the Nez Percé reservation in Idaho to help tribal members capture and record their old songs and stories before all of the old timers died off. Elizabeth Wilson, who had met Chief Joseph at her graduation from Carlisle Indian School in June of 1904, was still active at ninety-one years old and remembered many of those legends. And she was anxious to reveal them. After many days of recording, I returned to the lab and edited segments, extracting lines from those that spoke to the issue of noise and sound and reassembling them in a poetic narrative titled *Legend Days Are Over*. It is an example of the sentiment that pervaded almost every story she told and song she sang:

> The way the medicine men went and got guiding spirit
> Contact with animals or whatever it is,
> They kept on dancing every winter.
> They got strong and power came to them. Power came to them.
> Everything was different.
> It must have been in those times when everything was different
> Clear air and wilderness, and they could get in touch with animals
> like that.
> But I don't think they can now.
> Everything gone – noise and all …
> All right! Legend days will be over; humanity is coming soon.
> No more legend days.
> There will be no more
> And they will be sad like I am,
> Brokenhearted over my last child
> Never to return again.
> Death takes her. (Krause 2012)

## The Beaver's Lament

Nothing, however, stands out quite as much as this recent incident: truly, the saddest vocal expression I've ever heard came from a beaver.

During the spring of 2009, a fellow recordist from the Midwest sent me an audio clip of an event that took place at one of his favorite recording and listening sites – a remote pond in central Minnesota. No development surrounded it and there was no scientific or land management rationale for the heartbreaking event that took place that morning.

While recording one spring day, my colleague watched in stunned silence as a couple of game wardens appeared on the scene, planted some explosives, and blew up the beaver dam that, for years, had helped establish and maintain the subtle ecological balance of the habitat at the pond's outlet. Since there were no houses or nearby farmland to be protected, it seemed like an act of willful violence and unnecessary authority. The beaver family, its young and female, were decimated in an instant when the dam was blown apart. Remaining behind after the wardens left, my friend captured an altered habitat that no photo could have revealed. After dusk, the surviving and likely wounded male swam in slow circles around the pond, crying out inconsolably, its voice breaking in obvious pain as it searched for its mate and young family. Its vocalization is so forlorn and tragic that the recording is always emotionally difficult for me to hear or even speak about.

Although tail slaps and short vocalizations of beavers in and around their dams have been heard and even recorded on a few rare occasions, that's the first and only time I've heard beaver extended plaintive wails of this type. I hope never again to experience cries like those coming from another living being. The most heart-rending human music I've ever heard doesn't come close (Krause 2012).

### The Ba'Aka Pygmies

And finally, I want to end this section with an example from the Ba'Aka, a group of Pygmies who live in what was once the most remote part of the Central African Republic called the Dzanga-Sanga Rainforest, located in its southwest corner. Their habitat has been utterly transformed over the past thirty years because of both clear-cutting and poaching. Louis Sarno, an American ethnomusicologist, arrived in the Dzanga-Sangha region just before radical changes such as increased logging, poaching, missionary pressure, and other seductive lures drew members of the group into the cash economy. With the tribe not trusting him at first, he had to spend many months participating in the group's fraternity activities – for instance, eating bowls of live grubs given to him as one of many tests – until finally he was accepted as "Oka Amerikee," loosely translated as the "Listening American" (Sarno 2006).

When Sarno crossed the threshold of the Dzanga-Sanga, the link between the sounds of the forest and the music of the Ba'Aka seemed so strong that, he realized, without the biophony providing an obvious acoustic model of their spiritual and material world, their music would not have evolved as it did. Time and again, he witnessed his group break into performances, and the more he became familiar with both the music and the habitat, the more he recognized the mimetic connections between their music and the forest rhythms of the insects and frogs, the solo voices of birds, the occasional mammalian punctuation, and the multiple ways the sonic structures of the human and animal often reflected one another. Sarno's many accounts reveal how he came to hear the compelling spiritual, social, and practical connections between the biophonies of the forest and the resulting Ba'Aka music; the biophony was the equivalent of a lush, natural karaoke orchestra with which they performed (Sarno 2006).

The wild sonic environment that the group surrounded itself with comprised voices of their existence since their arrival in the Dzanga-Sangha. It may have been the primary beacon that lured them there in the first place. Describing a performance in an unpublished manuscript, Sarno (2006) wrote:

> This was *esime,* an extended rhythm sequence tacked on to the end of every song in a number of different dance forms, particularly those with drum accompaniment. What *esime* lacked in melody, it made up for in the complexity of its densely packed blocks of polyrhythm. Each woman had her own cry – a meaningless sound, a word, a rapidly uttered phrase – which she repeated in a characteristic periodicity. Each periodicity was unique. Some speeded up, broke their own beat, went retrograde. Tone hadn't lost all significance, either – in the array of cries and sudden conjunctions of two or more periodicities, intervals of minor seconds, of diminished and augmented sevenths, prevailed ... Two women elaborated on [these] constituent phrases with improvised

recitatives of yodel ornamentations and fanfare formulae, in a mind-bending display of free-flowing counterpoint that would have astonished the likes of Max Reger.

The soundscapes of these Central African forests are dazzling and blissful; the magnificent sonic cross-fertilization straddles multiple species. Lowland gorillas strike rhythms on their chests that are surprising and intricate. Forest elephants forage in marshy open meadows, bellowing low raspy growls – their sounds more felt by humans than heard – that reverberate over great distances. Black and white-tailed hornbills sail over the canopy, their raucous calls and the edge-tones of their beating wings subtly changing pitch as they find airborne purchase and pass in high arcs overhead. Goliath beetles hum and buzz. Red colobus and putty-nosed monkeys shout sforzando alarm calls to members of their groups. Hammerkops, ibis, and parrots pierce the air with their screams and calls. A wide variety of insects and frogs add a constant hum-and-buzz counterpoint to the acoustic fabric. One only needs to listen to Sarno's lustrous early recordings to hear the deep connection. Described by him as "one of the hidden glories of humanity" (Sarno 2006), Ba'Aka music emphasizes full, rich textures and bright-sounding harmonies. The sonorous qualities, intricate rhythms, consonance, and dissonance flow from and are influenced by their native biophonies.

But these days a change is occurring in the Dzanga-Sanga Rainforest. European Union, Chinese, and Japanese industrialists have discovered hardwoods and ways to extract them – albeit at a great cost to both the habitat and those who live there. Members of the group, seduced into Western money-exchange economies, have turned to prostitution and the poaching of bush meat to sustain their families, events that two decades ago would have been unheard of. That and the influence of media (via radio) and missionaries, whose sound is more like the lofty church and pop harmonies, have left indelible impressions on the music that the Ba'Aka once performed exclusively with forest biophonies as recently as the late 1990s. Since the tribe had historically relied on the sacred natural ambience of the forests around them, these changes have had a major impact on the physical and mental health of members of the group since their music has lost its original spiritual context. They now have to endure the humiliation of judgment as being "uncivilized" – both culturally and spiritually – as their new Western expansionist contacts constantly remind them. The social and physical pathologies inherent in the Western paradigms have begun to be transferred to the Ba'Aka. This physical and social upheaval causes the members to become ill. They become morose and quiet. Their skin breaks out in rashes and hives. They cannot eat and get respiratory infections as well as venereal diseases. When these illnesses become bad enough, the tribe will disappear for several months deep into the forest – much deeper than they traveled previously – to find some sense of solace and to heal from the effects of contact (Sarno 1996). The natural soundscape is the voice that helps revive them.

## The Wind in the Reeds

I want to end with a story of how the sounds of the natural world can inspire us to think differently. Lake Wallowa in northeastern Oregon is sacred to the Nez Percé. It was also the starting point for the 1877 flight led by Chief Joseph and several other chiefs, who, after outrunning and outfighting five American armies over the course of three months and a passage of seventeen hundred miles – with their entire tribes in tow, including children – were defeated at Bear Paw Battlefield in Montana, just forty-four miles short of the Canadian border and freedom.

When I first saw this idyllic site in October 1971, nestled at the foot of Chief Joseph Mountain in the Wallowa-Whitman National Forest, the shore was covered in a frosty mantle glistening in the early-morning sun. Guided there for a music lesson by Angus Wilson, a Nez Percé Elder whom a colleague and I had just met, we were steered to his sacred learning ground and told exactly where to sit. Then he moved some distance away while my friend and I waited, our interest piqued the evening before when Wilson alluded to the depths of our musical ignorance. Offering to reeducate us if we were at all curious, he cautioned that the tutorial might require a high degree of patience and a suspension of long-held beliefs. We were in.

Waiting by a feeder stream that flowed from a valley to the south, we squatted on our haunches, trembling – the temperature hovered in the mid-twenties (-5°C), and we lacked proper clothing to sit on the ground without cover. Anticipating an unnamed and unexpected event, we impatiently scanned the valley, where the woods were dense with lodgepole pine, Douglas fir, western larch, ponderosa, and low-lying scrub. Except for the sound of a few ravens, it was, at first, pretty quiet. Nothing moved. Wilson hadn't said much on the three-hour predawn journey from Lewiston, Idaho. When we got to our destination, he simply asked us to cool our heels and said that all would be exposed in time.

After about half an hour, the wind began to funnel down from the high southern pass, gaining more force with each passing moment. A Venturi effect caused the gusts passing upstream through the narrow gorge to compress into a vigorous breeze that swept past our crouched bodies, the combined temperature and windchill now making us decidedly uncomfortable. Then it happened. Sounds that seemed to come from a giant pipe organ suddenly engulfed us. The effect wasn't a chord exactly, but rather a combination of tones, sighs, and midrange groans that played off each other, sometimes setting strange beats into resonance as they nearly matched one another in pitch. At the same time they created complex harmonic overtones, augmented by reverberations coming off the lake and the surrounding mountains. At those moments the tone clusters, becoming quite loud, grew strangely dissonant and overwhelmed every other sensation.

While never unpleasant, the acoustic experience was disorienting. The sound came out of nowhere and completely masked the natural soundscape; we couldn't connect the reverberation with anything we could see. With mixed emotions, my colleague and I looked at each other, puzzled and a bit apprehensive. Neither of us had ever heard of or experienced anything like this – nor had we thought to record in a way that caught the whole event.

Some time slipped by, and Angus slowly got to his feet, walking with arthritic stiffness over to us. He asked if we knew where the sound was coming from. Bordering on hypothermia, we both shook our heads. Placing himself between my friend and me, Angus motioned for us to stand up and walk with him to the stream bank. He asked again if we had any idea what was happening. Only then did we realize what he wanted us to hear and see. In front of us stood a cluster of reeds of different length, which had been broken off by the force of the wind and weather over the course of seasons. As the air flowed past the reeds, those with open holes at the top were excited into oscillation, which created a great sound – a cross between a church organ and a colossal pan flute. With that realization came the instantaneous release of all the tension in my hypothermic shoulders. If it hadn't been for this moment, we never would have given a collection of reeds growing in a remote area by an Oregon lake a second thought.

Seeing recognition in our faces, Angus then took a knife from the sheath at his belt and walked into the shallows, boots and all. He selected and cut a length of reed from the patch, bored some holes and a notch into it, and began to play. After performing a short melody – one that we did manage to capture on tape despite the freezing temperatures – he turned to us and, in a slow, measured voice said: "Now you know where we got our music. And that's where you got yours, too." Humbled, I realized this was beyond a doubt my most memorable music lesson.

## Conclusion

Richard Louv, author of *Last Child in the Woods* (2008), posits that as we lose access to wild habitats and thus our connection to the natural world, we become impaired with what he has identified as Nature Deficit Disorder (NDD), a type of infirmity that introduces debilitating kinds of physical and emotional tension into our lives. The late Paul Shepard suggested that the more we put distance between ourselves and the natural world, the more pathological we become as a culture. (Look at us!)

I, too, have found that in my life, especially with the incredible rate of loss of wild soundscapes, I am often saddened when I think that young folks will never hear, in real time and in wild places, the lush biophonies I benefited from so much as a child and adult. As we hear these voices of the divine, we're endowed with a sense of place – the true story of the world in which we live. With their loss, the health of every contiguous biome as well as our own sense of well-being is affected, sometimes in profound ways since the deficit implies a conditional change for almost all the living organisms that inhabit them whether human or Other. The whisper of every leaf and creature implores us to embrace the natural sources of our lives that, indeed, may hold the secrets of love for all things, especially our own humanity.

I borrow an allusion from crickets:
their song is useless,
it serves no purpose
this sonorous scraping of wings
But without the indecipherable signal
transmitted from one to another
the night would not
(to crickets)
    be night.
*José Emilio Pacheco*

NOTES

1 Many of the stories found within this chapter, including the stories about the Wy'am, the beaver, the Ba'Aka, the Nez Percé, and the concepts of geophony, biophony, and anthropophony have appeared in *The Great Animal Orchestra* (2012). These stories have been rewritten and reframed for the purposes of this collection.

REFERENCES


Krause, Bernard L. 2012. *The Great Animal Orchestra: Finding the Origins of Music in the World's Wild Places*. New York: Little, Brown and Company.

Louv, Richard. 2008. *Last Child in the Woods: Saving Our Children from Nature-Deficit Disorder*. Chapel Hill: Algonquin Books.

Sarno, Louis. 1996. Bayaka: *The Extraordinary Music of the Babenzélé Pygmies*. Roslyn: Ellipsis Arts. Sound recording.

– 2006. Unpublished manuscript. Used with permission.

Schafer, R. Murray. 1977. *Tuning of the World*. New York: Knopf Press.

# 2 Environmental Mourning and the Religious Imagination

NANCY MENNING

## Introduction

We don't grieve abstractly; we mourn particular losses of people, places, animals, objects, and ideas to whom and to which we are attached. These losses range from the most direct to those grasped only in the historical imagination. The death of a pet, the paving over of a favourite childhood playsite, the roadside body of the deer we just hit with our car, river gorges obliterated by dam-building projects, mountaintops removed by mining, once-familiar landscapes permanently transformed by nuclear contamination, night stars obscured by light pollution, and the absence of buffalo moving freely across the Great Plains are all grievable losses for those who have intimately known what is now gone. Even Bill McKibben's (1989) notion of an "end of nature" – which refers to the demise of an *idea* of nature and not to any tangible natural entity – represents a loss that may be mourned.

Mourning rituals and practices respond to the grief we feel when we lose those we love and to whom we are attached, enabling us to go on in a world of ubiquitous change and repeated loss. We love and are attached to people, places, things, and ideas, and yet our most richly developed and culturally embedded mourning practices are those developed to mourn human deaths. The paucity of our rituals and practices to mourn ecological losses is troubling.

Ecological losses differ in important ways from human deaths. In particular, we are often complicit in these losses, if only by virtue of living in the Anthropocene. We must mourn not only what we have lost, but also what we have destroyed. While individual deaths and even extinctions occur naturally, most environmental losses are caused by human activity. When one feels complicit (directly or indirectly) in the loss being mourned, guilt entwines with sorrow, complicating the grieving process. This is especially true if we view ourselves as caregivers or trustees who are somehow responsible for the lives or well-being of those entrusted to us. These losses must be both mourned and repented.

Mourning rites developed within the diverse religions of the world, whether practised by a Jewish son saying Kaddish for his father over the course of nearly a year or by a Tibetan lama reading the *Bardo Thödol* to a deceased person's consciousness for forty-nine days, are historically constructed, conceptually articulated responses to loss. These religious ceremonies to mourn and memorialize human deaths serve as useful analogies for imagining rituals to mourn environmental losses. In this chapter, I explore Dagara funerals in West Africa, Jewish mourning practices, and Tibetan Buddhist death practices. The particularities of these diverse mourning rituals clarify our understanding of how rituals work in the context of their specific belief systems, helping us conceive culturally appropriate possibilities for mourning ecological losses. I offer a fourth example – Shiite Ashura rites – to illustrate a mourning practice capable of transforming the guilt resulting from our complicity.

We harm ourselves as we harm the world around us. Effective grieving must not only salve our loss but also assuage our guilt. The first three examples developed in this chapter – from Dagara, Jewish, and Tibetan Buddhist cultures – will begin to help us think creatively as we imagine rituals to mourn ecological losses. The fourth example – from Shiite Muslim culture – will help us address the special challenges of mourning losses that are complicated by guilt. I preface these four examples with a reflection on how religious rituals and spiritual practices respond to the challenges of loss, grief, and mourning and support sustained environmental activism.

## Religion, Mourning, and Environmental Activism

Denial shapes our actions (or inactions) in powerful ways. If I am in denial that my grandmother is dead, I am unlikely to be attending well to my mother's health. A similar logic underlies the concern raised by Douglas Burton-Christie (2011) when he suggests that our capacity to respond adequately to current and pending environmental challenges may be predicated upon having mourned properly the environmental losses we have already sustained. "Can there be genuine and lasting ecological renewal," he asks, "without a deep expression of grief and mourning for all that is being lost?" (29). Denial is merely one possibility; many of us are profoundly inattentive and largely unaware of the losses that surround us. Adequate mourning presupposes adequate connection and both are necessary to sustain environmental activism. Reflecting on a rare cutleaf silphium that bloomed in a graveyard west of Madison, Wisconsin, Aldo Leopold (1949) wrote: "We grieve only for what we know. The erasure of Silphium from western Dane County is no cause for grief if one knows it only as a name in a botany book" (48).

We must then cultivate our sense of connection, our feelings of grief, and our capacity to mourn. Such cultivation is an active process that may be nurtured and sustained by spiritual practices and religious rituals. No firm distinction is intended here between "spiritual" and "religious." The more important distinction is between practice and ritual. Practices such as daily prayer or meditation are intentional, repeated performances that are valued as ends in themselves. They are performed (practised) on a regular basis as part of a spiritual discipline. Rituals, on the other hand, are generally performed to achieve some external end; they are means to that end rather than ends in themselves. They are performed as needs arise. For example, daily meditations on death are a practice; the funeral performed on the occasion of a specific death is a ritual. That said, the distinction between practice and ritual is not rigid; practices and rituals exist along a continuum, not in distinct categories. Both practices and rituals may nurture our sense of connection, work with our feelings of grief, and further our mourning process.[1]

Religious rituals are structured activities, moving those who grieve through their mourning. With respect to death and other kinds of losses, the primary rituals of interest are rites of passage: patterned actions that focus our attention on important transitions. As theorized by Arnold van Gennep (1960) and Victor Turner (1969), rites of passage follow a three-stage process of separation, liminality, and reincorporation. The loss itself brings about the separation. The grief that arises is one manifestation of the personal and social – and, in the case of environmental losses, ecological – disruptions characteristic of the liminal (or transitional) period. Mourning rituals structure activities and interactions during these bewildering and disorienting periods, helping survivors reorient their lives in the absence of what has been lost. There are two basic paths forward: we might seek to maintain our bonds of connection with that which has been lost to us, even in its absence, or we might seek to replace the lost person or place or object with another person, place, or object that is available to us.[2] While memories remain and grief may continue in some form, reincorporation marks the time when one's personal identity, social relationships, and ecological interactions have been largely refashioned in the wake of the loss, thereby concluding the formal rite of passage. Healing, then, is not a restoration of the way things were; rather, it is a reorientation to a relatively secure and stable life in the absence of what once was. Mourning rituals nurture, if not ensure, this healing.

The stereotyped patterns of ritual activity are often welcome to the extent that those suffering a loss are in need of some help, some consolation provided by custom and tradition. This is especially helpful in liminal situations. Connectedness is our natural condition. We are social beings. And that sociability extends to biophilic, topophilic, and cosmophilic dimensions. Severed connections at any of these levels propel us into unsettled, liminal conditions that are profoundly disorienting. Ritual, when performed collectively, draws those who are less profoundly wrecked by the storms of loss around those who are bereft, nurturing and supporting them as they put the pieces back together again, building a new structure and orientation for their lives after loss.

Ritual is a powerful force, performing important work in the face of profound loss. It quells fight, flight, and freeze responses, directing our attention to small actions and sequences of actions that we take individually and collectively in disorienting times. It marks the loss, directing the community's attention to the significance of the loss. It provides a temporal, spatial, and social framework upon which to begin reorienting what has become disoriented. And it has an end. We have not just been thrown into the dangerous waters of liminality. Rather, our attention, again both individually and collectively, is on moving through a dangerous (and powerful) space-time to the farther shore of renewed stability and connection, wholeness and ease.

While rituals help us respond well to particular losses, practices may help us become the kind of people who are resilient in the face of impermanence. We suffer repeated losses. The world is always falling away. We need practices, ways of meditating on death. Beyond the funeral and memorializing rituals that mourn particular losses, some religious traditions describe disciplined spiritual practices that can transform the way we understand, prepare for, and respond to death. Much needs to be brought to attention: our understandings of the world we live in, what is important, who we are as humans, the transience of our relationships, the pain we feel, our own impending end. Ritual produces that kind of attentiveness and practices nurture it. Meditative practices may also draw our attention to the continual strengthening of bonds that remain but which fray over time and to the intentionality of establishing new bonds that are conducive to ongoing wholeness. When done well, these spiritual practices make us more present to ourselves and more attentive to all that we love, thereby deepening our experiences and enriching our lives.

Even if we recognize the need to mourn our losses, it is not always obvious which practice or ritual to perform. We need help, then, to imagine effective mourning rituals for the environmental losses we suffer. Phyllis Windle (1992) argues that biologists and ecologists need rituals to mourn the seemingly inevitable losses of organisms and field sites to which they have become personally attached. But she then asks: "Even if I decided to grieve, how would I go about doing it?" (365).

While losses are pervasive and manifold, arising in all aspects of our lives, our best-developed mourning practices are in response to human deaths. The religious imagination offers one means of making the translation from human deaths to environmental losses. Indeed, the religious imagination is a critical thinking skill anyone may develop, regardless of philosophical or religious commitments. In broad outline, it is a capacity to interpret generously the particularities of rituals performed in distinct cultural and historical contexts, to reflect thoughtfully on one's own personal and cultural commitments in dialogue with the beliefs and practices of others, to apply creatively the insights derived from one context (human deaths) to another context (natural or environmental losses), and to commit vulnerably to newly imagined practices and rituals for mourning environmental losses (and sustaining environmental activism) aligned with the circumstances of one's own biographical, ecological, and historical context. By drawing careful analogies from human deaths to ecological losses, we can employ the religious imagination to guide our creation and implementation of effective mourning practices.

Losses are specific and the rituals constructed to mourn them must be suitably matched to the specificity of those losses. We are interested in the ways rituals reinscribe connections, guide emotions, and move mourners through a liminal period. Due to the anthropogenic nature of many environmental losses, we will also need to consider the ability of mourning rituals to address and transform feelings of guilt and remorse. Finally, we will want to evaluate whether performance of the rituals or practices deepens our commitments to ongoing environmental activism that might forestall, mitigate, or prevent future losses.

## Religious Models of Mourning

The death of a human is a profound loss and religious traditions have shaped and transmitted rituals to assist in acknowledging connection, honouring loss, and dealing compassionately with grief. These rituals are socially effective and culturally meaningful ways of marking loss, easing pain, and rebuilding disrupted lives. Mourning rites, especially in their social aspects, are culturally embedded – they draw their meaning and their effectiveness from their contexts. Each culture emphasizes certain aspects, drawing out one element rather than another. If we can begin to understand differences between rituals, we may begin to see our own context more clearly and develop insight into what might make for effective mourning rituals for our particular ecological losses.

I selected the rituals and practices presented here for several reasons. First, I chose rituals that are well documented in readings that are accessible to a wide variety of audiences. Second, I selected rituals that present diverse and divergent approaches to death. Third, and most importantly, I chose rituals that have distinctive elements that invite thoughtful reflection on the challenges of environmental mourning.[3]

## Dagara Funeral Traditions

The Dagara are an Indigenous people of West Africa with traditional homelands in the savannahs of Ghana and Burkina Faso. Jack Goody has been the dominant ethnographer of the Dagara, contributing significant insights into Dagara culture through his work with the traditional myths and initiation rituals employed by voluntary associations (e.g. Goody 1972; Goody and Gandah 2002). Dagara funerals – the most elaborate ceremonies of the Dagara, "whether this be measured in terms of numbers attending, time taken, or emotion generated" (Goody 1962, 11) – have been relatively resistant to change despite cultural discontinuities due to the introduction of Christianity and formal education (Alenuma 2002; Kuukure 1985). Goody's work on funeral customs is included in his classic book *Death, Property, and the Ancestors* (Goody 1962). Another useful resource is the writings of Malidoma Patrice Somé, a Dagara shaman who has sought to promote the importance of ritual to Western audiences (Somé 1993, 1994, 1998).

Imagine this scene at a Dagara funeral (compiled from details in Goody 1962 and Somé 1993). Two women elders sit alongside a corpse, which is propped up on a shrine built from ebony poles and guinea-corn stalks, occasionally whisking the flies away. A crowd has gathered. For most, attendance at a funeral is compulsory, whether they have been summoned by the ululation of women in their local village, by the sound of the *gyil* – the Dagaran national instrument similar to a Western xylophone – from a nearby village, or by a messenger sent from a more distant location. *Gyile* play continuously. Into the empty space between the corpse's shrine and the gathered community – a sacred space representing turmoil and chaos – run mourners, crying lamentations; their hands are clasped behind their necks in a culturally recognized symbol of grief. Mourners most affected by a death – parents, spouse, siblings, and children – are restrained by ties around their waist, wrist, or ankle, which are held by companions who shadow their movements, ever watchful that they not commit suicide in the midst of their sorrows or have their souls abducted by the still-present soul of the deceased.

Emotion is central to the Dagara funeral ritual. "When death occurs, the thing to think about is not burial, but funeral, tears, grief" (Somé 1993, 84). The Dagara understand death as that which "stands between the world of human beings and the world of the spirits, between the visible and the invisible" (Adeyemo 1979, 64). Intense emotion is needed to cross this boundary. "Tears carry the dead home" (Somé 1993, 75). For the soul of the deceased to make the transition successfully from the Land of the Living to the Land of the Dead, both spirits and people must help. "The spirits are invoked mainly so that they can come to help the deceased in the journey ahead by squeezing enough emotion out of the hearts of the grievers" (ibid., 76). While designated companions must restrain primary mourners from actions endangering their own lives, Dagara funerals orchestrate the effusive outpouring of grief. "The more one grieves, the more one gives to the dead and the more one moves closer to being with the dead. The container's task is to ensure that the mourner knows well how to distinguish between a grief that helps the dead soul go home and a grief that kills" (ibid., 81–2). The elders who sit alongside the body are "accompanying the dead person, collecting all the grief poured into the space and loading it on the soul of the dead one as it readies itself for the grand departure" (ibid., 77).

The Dagara culture's emphasis on tears and grief recalls Douglas Burton-Christie's (2011) insight that weeping nurtures intimacy. Tears "open up the soul to an honest and deep reckoning with who one really is – in relation to everyone and everything – and in the process lead toward a reknitting of the torn whole" (35). Dagara funeral practices are illuminating for the way they both give expression to our vulnerability and orchestrate a practice to heal our relationality. After the emotional outpouring of grief has run its course, friends of the deceased person make speeches, relating how they came to know the deceased, and make offerings to the dead, which are accepted by an immediate relative of the deceased – thereby transferring the bond of friendship. Henceforth, the new friends are bound by the same obligations as the old (Goody 1962). In the midst of chaos and the disruption of daily life, friendship transfers reweave the social connections severed by death. The exception is with the death of a parent; a parent cannot be replaced.

Ritual actions reveal the relationships disrupted by the death and the implications of the death in a broader social context. Funeral ceremonies for an infant may last only a day, while those for an elder may continue for three to five days. At the conclusion of the funeral ceremony, gravediggers take the corpse away to be buried. Most are buried in the bush. Elders, however, may be buried in the family compound; when in need, living family members offer libations on the shrine, and the deceased elder is able to continue to care for the living.

As a model for environmental mourning, a distinctive aspect of the Dagara ceremony is full participation by everyone in the community. Losses are differentiated based on the depth of social connections and whether a relationship can be replaced. The corpse is present while the ceremony is in process. Grief is openly and effusively expressed, though within safe bounds. This example compels us to ask: what ecological losses merit mourning by entire communities? For any particular loss, who are the primary mourners, who are the containers, and who are the other members of the community? How can we match the ecological significance of the lost environmental feature to the length of time a mourning ritual extends? Which ecological losses constitute the equivalent of deceased "elders" who must be enshrined so that they remain ritually present into the future? For whom is any particular loss irreplaceable, and to what alternative environmental feature might others shift their ecological relationships? How can we keep the lost entity present to the mourning community during the mourning ritual? And how can emotional expression be encouraged to a safe yet cathartic level?

## Jewish Mourning Practices

Traditional Judaism is a Talmudic or rabbinic Judaism, looking back for guidance to the canonical texts written by Jewish leaders in the centuries following destruction of the Second Temple in Jerusalem in 70 CE. Lawrence Hoffman (1996; see also Goldenberg 1992) provides a comprehensive account of traditional Jewish mourning practices. Numerous books published in recent decades offer detailed guidance for performing mourning rituals (e.g. Diamant 1999; Lamm 1969).

Rabbinic tradition defines the mourners by familial relationships: parents, siblings, spouses, and children mourn the deceased.[4] Mourning does not commence until the corpse has been buried. Upon hearing of the death, grief-stricken mourners are plunged into a liminal period, rending their clothes to symbolize their bereavement. Once the corpse has been properly washed and buried, proper mourning begins with a seven-day period of sitting shiva. Mourners remain home from work and are visited by friends and more distant relatives who honour the dead by bringing meals and other support to the mourners. Mourners return to work after a week and, in structured and well-defined stages, continue to transition back into normal participation in public and ritual life over the course of the following year. A central component of the yearlong mourning is daily recitation of the Kaddish – a traditional, formally structured mourner's prayer.

Ritual prescriptions are not ritual performances. Traditional guidelines for the ritual requirements of Jewish mourning do not reflect actual Jewish practices by Jews of distinct and divergent traditions (Reform, Orthodox, etc.) with different and ongoing adaptations to modernity. Hoffman (1996) notes, for example, that Reform Jews still sit shiva, but often for only three of the ritually prescribed seven days. Recent memoirs by sons saying Kaddish for their fathers provide important insights into actual Jewish mourning practices (Goldman 2006; Heilman 2001; Wieseltier 1998). Specifically, these works caution against an overemphasis on the doctrinal reasons undergirding ritual practices (at least within Jewish tradition) and draw our attention to the important work done by ritual in reconnecting the bereaved to the broader community.

Kaddish practice reweaves the bereaved child into the enduring social bonds of community, thereby placating the irreplaceable loss of a parent. "The obligation to say kaddish ... thrusts the mourner out of his or her home and into the community at a time when it might be easier to withdraw and quietly grieve" (Goldman 2006, 60). Daily, for eleven months after the death of their parent, an adult child (most often a son) is obligated to recite the Kaddish prayer in a minyan, the quorum of ten required for Jewish communal prayer (Goldenberg 1992, Hoffman 1996, Lamm 1969). Others – daughters, additional sons – may voluntarily adopt the practice as well.[5]

Practices matter more than beliefs here. The Kaddish, an ancient Aramaic prose poem, begins, "Magnified and sanctified / may His great Name be / in the world that He created" (Wieseltier 1998, xiii). Beyond expressing a general eschatological hope for peace, the mourner's Kaddish says nothing about death or grief; reference to an afterlife is "conspicuously absent" (Diamant 1999, 20). The eleven-month span of Kaddish practice, admittedly, may be explained by reference to the afterlife, in that the Kaddish prayers of children are said to lessen the punishments suffered by their parents in *gehinnom* (hell) before going on to *gan eden* (paradise), but a practising Jew can readily set that aside. "I had heard the rabbinic explanation many times before, but it was not something I wanted to think about in the context of my mother and father. So I just followed the rules" (Goldman 2006, 192). Judaism, a more orthopraxic than orthodoxic tradition – i.e. asserting the primacy of correct practice over correct belief – lacks a doctrinal view of the afterlife, and individual Jews hold diverse beliefs about what happens to humans after death (Diamant 1999; Goldenberg 1992).

Thus God is praised in the midst of tragedy in a prayer practice that turns the practitioner – "shielded from a fatherless world by a fatherful practice" (Wieseltier 1998, 455) – away from death back toward life, away from isolation and back toward community. "The recitation of Kaddish has united the generations in a vertical chain, father to son, while the requirement to gather a minyan for Kaddish has united Jews on a horizontal plane" (Lamm 1969, 157). Slowly, intentionally, the practice weaves the bereaved into an encompassing narrative of faithfulness. Writing partway through his year of Kaddish, memoirist Leon Wieseltier (1998) realizes he has been "thoroughly absorbed into the fellowship of these excellent people. I am no longer a stranger in my community" (273). At his father's gravesite, on the first anniversary of his death, he looks out upon the old men, "huddled against the wind," who had come to pray one more time for his father; "they are getting to their end, I thought; and I loved them; and I wept" (584). Wieseltier and other Jews writing memoirs of their Kaddish practices are often not particularly observant Jews in the years preceding the death. Nonetheless, their worlds rocked by the deaths of their fathers, they turn to tradition to take up the generational responsibility of carrying the tradition forward. In so doing, they are reconnected to a sustaining community.

Whereas in the Dagara funeral ceremony the body was kept present while the gathered community grieved, in the Jewish tradition mourning begins only after the body has been properly buried. The full Dagara community mourned a death; in Jewish practice, most of the community is unaffected by a loss in any substantial way, while immediate family members are allowed to re-enter social, cultural, and economic life over a defined, extended time period. And whereas the Dagara funeral ceremony withheld the potential of friendship transfer when a parent died,

Jewish tradition offers a lengthy practice – traditionally incumbent upon the son of a deceased parent – that binds that son to the community, thereby effecting a generational transfer of cultural responsibilities. We might ask: for environmental losses, when is it appropriate to put the loss to rest before entering into a mourning process? For those particularly affected by a particular ecological loss, how can the broader society support them while they transition slowly back into full participation in ongoing community life? Finally, how can we connect those who mourn a particular loss with others who mourn other losses in a substantial ritual practice that sustains broader cultural connections? Who might commit to joining with those who mourn in ritual practice – forming a minyan, as it were – even though they have not themselves suffered any recent loss?

## Tibetan Buddhist Death Practices

Tibet's isolation lends itself to mythic status (see Thubron 2011), complicating our efforts to accurately understand and respectfully interpret Tibetan Buddhist death practices. For example, Tibetan Buddhism, imagined through texts and practices, has tremendous appeal in the West. The popular English translation of the *Bardo Thödol* (the Tibetan Book of the Dead) by W.Y. Evans-Wentz in 1927 offered Western audiences rich images of experiences in the forty-nine-day liminal period between death and rebirth in the ongoing cycle of reincarnation (Karma-glin-pa 2000). Though its "romantic Orientalism" is "fraught with problems," the *Bardo Thödol* nevertheless remains popular (see foreword by Donald Lopez in Karma-glin-pa 2000, B, G). And it remains influential, retaining its usefulness even among American Buddhists for whom Buddhist ideas of reincarnation are largely irrelevant to practice (Goss and Klass 2006). Tibetan Buddhist sky burial practices, in which corpses are fed to vultures, fascinate Western audiences due to such close alignment with nature at its most elemental (Faison 1999, Kelly 1997), offering images of death conducive to poetic interpretation (Batt 1988, Levin 2011).

While sky burial can be explained by environmental limitations, the practice also transmits important religious lessons. On the high Tibetan plateau, trees to fuel cremations are scarce and the ground is often too rocky or frozen for burial. In this setting, some corpses are carried to a ritual site, the flesh flayed from the bone, and the body consumed by vultures. *Jhator*, the Tibetan term for this burial practice, means "giving alms to the birds" (Camphausen 2013, n.p.). This practice reinforces core Buddhist beliefs about reincarnation and liberation. Tibetan Buddhists seek liberation from the endless cycle of reincarnation. Our ability to achieve liberation is enhanced if we are able to maintain a clear and focused consciousness during this liminal period that follows our death. Grasping at the materiality of our bodies hinders liberation. "Tibetan Buddhists are encouraged to observe *jhator* in order to confront the realities of physical death without fear. After all, for them, the real trials of death are inner ones, while the fate of the outer body is a mere passing on of nutrients to other beings" (Lamb 2011, n.p.).

In a sky burial, ritual dismemberment of the corpse is a means of "break[ing] physical attachments of the relatives to the deceased" (Goss and Klass 1997, 385). Nonetheless, while death provides an opportunity for Buddhists to release emotional ties to the physical bodies of their loved ones, mourners maintain their bonds with the deceased by "channel[ing] their grieving energies into spiritual practices for the dead" (Goss and Klass 1997, 389). Part of the survivor's new life story becomes to transfer merit to the deceased one's consciousness, a work that will endure with practices marking every anniversary of the death. Sky burial makes sense against a cultural backdrop of beliefs in karma – "*karma* being the addictive desire to remain human, to rest in an illusory world of phenomena which grows stronger with every reincarnation" (Grainger 1998, 15). The notion of karma also includes belief in the pervasiveness of impermanence and suffering and in the possibility of enlightenment, understood as release from the otherwise endless cycle of reincarnation. Attachment to a deceased body – in a mountainous landscape that resists decay, no less – hinders this liberation project. The physical body, no longer inhabited by consciousness, may be offered to others – to the vultures. This "giving alms to the birds" is an act of compassion that accrues merit, fostering enlightenment.

Sky burial stimulates deep reflection. "Tibetans say that everyone should witness a sky funeral at least once. It brings home the transience and impermanence of life, urging us not to waste the precious opportunity afforded by a human life" (Kelly 1997, 83). In this spirit, poets have imagined sky burial practices from the perspectives of the deceased and of those who look on and who participate in the ritual dismemberment. In his poem "Sky Burial," Herb Batt (1988) imagines the experience of a deceased woman. Her brothers carry her to the ritual site, where her mother and son wail to see her body on the cutters' stone. A lama reads to her consciousness from the *Bardo Thödol*, guiding her to enlightenment or to a positive rebirth. Then the cutters begin. "The first takes off my skin / As if he drew my clothes off for the night. / His fellow hacks away my arms and legs. / Another slices off my head, / Then cracks my skull and scatters out my brains / Into the dust" (111). The vultures, alerted by an aromatic signal fire they know well, descend to the feast. Dana Levin (2011), working from the other side of the experience, writes from the perspective of a witness: "she is the jellied wealth they dive for, heads featherless and slick – / He's waving them away with a long white stick, your ritual-butcher friend. / You must finish, you must sever her head – / What is an arm but an alm" (9).

For Buddhists, the sky burial ritual is embedded in cultural practices of meditation on death that prepare practitioners for the inevitability of loss. To respond well when death arrives we must prepare ourselves for it. A classic text of Buddhist practice identifies ten objects of meditation centred on the corpse in various stages of mutilation and decomposition (Buddhaghosa 1976, Shaw 2009). When the *Bardo Thödol* is read over the course of forty-nine days to the consciousness of a deceased person, family members attending this reading absorb its perspectives on death, the material body, consciousness, and liberation (Sambhava 1993, *The Tibetan Book of the Dead* 2009). The practice of meditating on death complements the religious rituals performed once death has occurred. Thinking about loss and impermanence – about death – prepares us to mourn well when losses occur.

More so than the Dagara and Jewish examples, the Tibetan Buddhist death practices – with their adaptation to the environmental conditions of the high Tibetan plateau and the recognition of the cycle of life implicit in the feeding of vultures – invite ecological reflection. Nonetheless, as a model for ecological mourning some aspects of Tibetan practice present distinct challenges to environmentalist thought. In particular, the forceful lessons this tradition delivers about non-attachment to the physical body should provoke deep introspection on our attachments to natural entities in a world shot through with impermanence. The Tibetan Buddhist practice of meditating on death as well as the ritual performance of sitting with the corpse while the *Bardo Thödol* is read are models for this sustained reflection on death, loss, and impermanence. These Tibetan cultural practices contrast sharply with the Jewish practices of burying the corpse before beginning the mourning process and committing to a yearlong practice of reciting a Kaddish prayer that only obliquely references death. We ask: how can our mourning practices align with ecological insights as well as condition our responses to the inevitability of loss? What kind of losses call for dramatic ritual actions to break our unproductive attachments to what once was? How might practices of meditating on environmental losses help us maintain a useful equanimity when future losses occur?

## Shiite Ashura Rites

The three previous examples – drawn from Indigenous, Western, and Eastern cultural settings – already offer rich resources for reflection and for practice of the religious imagination. I provide a fourth example, Shiite Ashura rites, to illustrate a mourning practice capable of transforming complicity. We are often directly or indirectly complicit in the ecological losses we suffer. This final example will help us imagine mourning and memorializing rituals that may assuage our guilt, turning these energies toward more productive ends.

Husayn ibn Ali was the grandson of the prophet Muhammad. In 680, his attempt to assert leadership over the Muslim community met with disaster. The supporting army he had hoped for from the nearby community of Kufa, a stronghold of his father Ali, never arrived. Without their aid, Imam Husayn's small encampment was besieged for ten days and finally massacred on the holy day of Ashura, a day set aside half a century earlier by Muhammad for fasting and atonement. While Muhammad's Ashura practice was intended for the entire Muslim community, following Husayn's death, the Shiite Ashura ritual began to diverge, taking very different shape over the succeeding centuries. Just four years after Husayn's martyrdom, those who felt complicit in his death due to their absence on the Karbala battlefield engaged in a death march, perhaps seeking their own martyrdom. Far from seeking to break ties to the deceased, Ashura (sometimes called Muharram) was becoming a mourning ritual of deepening attachments.

Strengthening their ties with the martyred Husayn and with one another, mourners continued to mark his death. In the historical arc traced by Mahmoud Ayoub (1978) and Yitzhak Nakash (1993), each anniversary added a new page to a life story seeking to make sense of his loss and their lives. Initially Shiites, then a minority community, met secretly for private memorial services to mourn and recount the events of the battle in which they had failed their leader. Following preexisting mourning customs, they would visit the tomb of the dead on the anniversary of his death and again forty days later. During the Abbasid period (750–1258), private memorial services moved to public mosques and took on increasingly elaborate forms involving wailing and poetic narration. Centuries later, as Shiite-majority governments arose in parts of the Islamic empire, those governments took a more active role in promoting, sponsoring, and structuring Ashura rites. They created a carnival-type play to reenact the battle at Karbala; over time that play has become increasingly theatrical, complete with drums, tambourines, burning incense, street food, costumes, and horses (Afary 2003). According to Nakash (1993), self-flagellation was a relatively recent addition to the rite. Now, in Ashura's fully developed ten-day ritual mourning complex, private memorial services erupt into public spaces; the grieving walk in procession behind a symbolic coffin, weeping, beating their breasts, and slapping their own faces. The deadly battle is reenacted and its story of betrayal recounted. The tomb of the dead is visited. Some mourners beat themselves across the back and shoulders with chains, some using chains tipped in razors; others draw their own blood with knives and swords (Afary 2003; Nakash 1993).

"Ashura is not merely a commemoration of the death of a leader. It is a ceremony of atonement for sins, individual and communal" (Sindawi 2001, 205). The people's guilt in failing to come to Husayn's aid is inextricably linked to his martyrdom. By reliving the event as participants in ritual, his supporters join together to transform what happened – not the fact of what happened, but its meaning. Their historical absence in the battle at Karbala is redeemed by their ritual presence now. Though deeply complicit in Husayn's death, participants promise never to be absent again, vowing to henceforward be present when God's will is to be done or when social injustice must be confronted. "Every day is Ashura and every land is Karbala" (Norton 2005, 145). This activist orientation was a long time in coming, arising more than a millennium after the founding event. Lara Deeb (2005) analyzes how the meaning of Ashura shifted from regret to activism in Lebanon beginning in the late 1960s. Pious behaviour, for both men and women, became tied to sacrifice. Instead of the self-harm of flagellation, modern Shiite Muslims in Lebanon donate blood to local bloodbanks to commemorate Ashura. These late developments in Ashura ritual practices suggest how guilt over past failures can be transformed into a commitment to present and future activism while also illustrating the openness of established rituals to different and reimagined actions, better aligned with more contemporary and desired outcomes.

Similar to Buddhist meditations on death, Shiite Ashura rites keep the details of loss in clear focus. However, whereas the Tibetan Buddhist sky burial practice strives to break attachments to the physical body of the deceased, Shiite Ashura rituals deepen attachments to Husayn and the defining events of his death. The past must not merely be forgotten when survivors bear responsibility for the loss. Rather, the Shiite Ashura rite provides a model for a repentant and redemptive rite that transforms the guilt associated with past inaction into explicit activism to meet present and future ecological needs. This ritual example compels us to ask: how are we responsible directly and indirectly for environmental losses? How might we enact in ritual a different way of responding to environmental challenges in the past? How might the energy of this reflection, repentance, and ritual action be directed toward environmental activism in the present and future? How can the losses we've sustained – and even contributed to – become defining moments for our communities?

## Exercising the Religious Imagination

Losses are particular and the practices meant to prepare for them and the rituals enacted to respond to them must be thoughtfully matched to the specific loss. Each ritual considered here lends itself well to particular kinds of losses. The Dagara ritual model, for example, is fruitful for situations in which it is important to gather an entire community (perhaps the sudden devastation of a stretch of river due to a chemical spill) or situations in which a friendship transfer is appropriate (perhaps the replacement of one backyard for another when a family moves). Jewish mourning practices are helpful when the survivors of a tragic loss need some structured time to transition back to full participation in community life (perhaps during the recovery from a devastating natural disaster) or when the equivalent of a generational transition is imagined (perhaps when a power company cuts down a community's trees to protect its power lines from blowdowns and a newly dedicated park needs to be planned and brought into being). Tibetan Buddhist practices are useful models when we need a dramatic image or experience to break our unhealthy attachments to what has been lost (perhaps when an activist group's attachment to a past land use violation prevents their effective participation in negotiations about current and pending land use activities) or when we can develop clarity of mind and resilience by meditating deeply on future losses (perhaps as we imagine the likely multitude of losses and tragedies that will accompany continued climate change). And Shiite Ashura rites are compelling analogies when guilt complicates our grief. We may be complicit in many of the environmental losses that we wish to grieve – whether directly, if we hit a deer with our car, or indirectly, when we buy food grown with pesticides. Indeed, if we trace honestly the wideranging environmental consequences of our unbounded human-induced impacts, we may be overwhelmed with guilt at the volume of destruction engendered by our daily choices. The Ashura rites show that powerful rituals can arise in the conjunction of practices for both mourning and atonement. Out of loss, we can develop rituals that transform our feelings of guilt, redeem our past failures, and provide powerful motivation for transformative action.

Of course, none of these ritual models can be simply applied whole cloth to mourn environmental losses. Their utility lies in expanding our imaginative capacity as we find ourselves thrown into grief by some new environmental loss. Beyond imagining possibilities, critical thinking also demands careful attention to the limitations of each ritual or practice as a model for mourning environmental losses. Each of the rituals outlined here has limitations that must be considered; nonetheless, we can use the Ashura ritual to illustrate the kinds of questions that arise during this stage of critical thought. First, the martyrdom of Husayn is the defining historical event for the entire Shiite community; it is, in many ways, the event around which the community coalesces as having a distinct identity. Which environmental losses might play such a pivotal role for any particular community? Second, some might argue that the ritual sacrifices of Ashura substitute for real self-sacrifice, in the sense that what one has done in ritual need not be repeated in actuality. Might memorializing complicity for anthropogenic environmental losses actually hinder effective environmental action? Third, the battle at Karbala was a political battle and the subsequent rhetoric of Ashura portrays Shiites as oppressed by Sunnis. For environmental losses, what enemy (analogous to the Sunnis) need be evoked? Finally, the events at Karbala were a distinct historical event, happening at a discrete moment in time. How can one mourn the often more diffuse cascade of environmental degradations? How do we transition from particular losses to the myriad ongoing losses in our lifetimes? Reflection on these sorts of questions will help us make wise and effective choices as we translate practices for mourning human deaths to the challenges of mourning environmental losses.

As we turn to committing ourselves to rituals and practices appropriate not only to our distinctive losses but also to our cultural and historical contexts, the exercise of the religious imagination as a form of critical thinking requires us to evaluate alternatives. Continuing our focus on the Ashura rite, we can illustrate this critical reflection by contrasting Ashura with the Altars of Extinction project (Gomes 2009). More species go extinct each year than any human could name, know intimately, or grieve individually. We may need a ritual practice that interprets an overall interconnection of losses, the way the Altars of Extinction honours individual species extinctions to bring attention to an overarching mass extinction event. Nonetheless, the Shiite Ashura model presents distinct advantages over aspects of the Altars of Extinction model that diffuse our culpability and alienate us from our humanity. The Altars of Extinction project is a series of art installations that act as ritual spaces intended to help viewer-participants "collectively contemplate and grieve the extinction of plant, animal, and fungal species at human hands" (ibid., 248). One purpose of the project is to raise awareness of the current mass extinction event. "When faced with loss of such magnitude, we often shut down our capacity to feel ... in doing so we also diminish our responsiveness, our ability to let pain move through us to appropriate and creative action" (ibid., 248–9). The project seeks to bring people out of denial about the sixth mass extinction event we are now undergoing, but it does not elicit repentance or seek atonement. Participants mourn losses "at human hands" (ibid., 246, 248) and honour "those who have passed" (250) or who "have left this world" (246) without experiencing the extinctions as betrayals or massacres committed by the participant. In contrast, the participants in Ashura rites return imaginatively to the scene of the transgression and become present to that original and timeless event, both taking on their responsibility for the original loss and transforming their complicity into a commitment to live differently from this moment forward. The annual rites memorialize their complicity, offering recurring opportunities for repentance and redemption. Emerging out of these reflections, we might wonder how to incorporate the repentant aspects of Ashura into a ritual that is otherwise shaped on the Altars of Extinction model. Such an adaptation would be a fitting outcome of the exercise of the religious imagination.

## Conclusion

Grieving starts with loss. We sever some relationships ourselves. Other times, other agents or the brute vicissitudes of life unmoor us from that which sustains us. Loss is important to define precisely because mourning must respond in measure to the loss sustained. Loss points us back to connection. We grieve what we love, those entities with which we have relationships that are partly constitutive of our identities. While the emotions of grief are largely passive, coming upon us in response to our losses, we can cultivate our capacity for grief by practices that create, nurture, or draw our attention to connections. Burton-Christie (2011) calls for spiritual practices that more deeply embed us in our natural world; for example, we might cultivate tears in response to past, present, and imagined losses, drawing our attention to the connections we do have to the natural world, despite our persistent inattention. If grief is a response to loss, mourning is what we do with it. Mourning is the action we embark upon to heal what has been harmed within us by the severing of relationships that matter deeply to us. It is the structured way we go about holding and acknowledging and working with our feelings of loss and the concomitant maelstrom of grief.

While losses are pervasive and manifold, arising in all aspects of our lives, our best-developed mourning practices are in response to human deaths. In recent years, scholars and activists have sought and created rituals to deepen the experience of "knowing" our ecological losses, prompting deep grieving and effective mourning. Burton-Christie (2011) argues that to the extent that we remain ignorant or in a posture of active denial of emotional responses to the ecological losses we have already sustained – and which continue – we may be rendered impotent to actively address pressing environmental challenges. To mourn adequately, we need to face, "directly and without evasion, the harsh ambiguity of existence itself, including all that is broken and in need of repair – in oneself and in the world" (35). By grieving well, we transform our relationship to the natural world and ground "the work of authentic and sustained ecological renewal" (31).

Reflecting on distinctive rituals to mourn human deaths drawn from diverse religious traditions helps us clarify our own beliefs and commitments, and in turn helps us imagine potential rituals to mourn more-than-human losses. The particular examples of mourning and memorializing rituals explored herein – Dagara funerals, Jewish Kaddish practices, Tibetan Buddhist sky burials, and Shiite Ashura rites – move our understanding of ritual process and function well beyond the popular five stages of grief model (Kübler-Ross 1969, Kübler-Ross and Kessler 2005, Konigsberg 2011). Drawing on the religious imagination, we may critically analyze a given ritual and its context, thereby coming to a deeper understanding of our own cultural framework. We may then creatively imagine possibilities for applying these developing insights to the challenges of effective ecological mourning and to the sustained environmental activism such mourning supports.

## NOTES

1 In recent years, scholars and activists have described specific practices and rituals intended to help us connect, grieve, and mourn adequately. Burton-Christie (2011) advocates contemplative spiritual practices such as

"*memento mori*, a sustained reflection on mortality, not only [one's] own but also that of every living thing" (34). Timothy Hessel-Robinson (2012) encourages liturgical practices of ecological lamentation that affirm the goodness of the natural world, identify threats to ecological integrity, offer expressions of grief in the face of natural losses, dedicate the participants to solidarity with the suffering of the more-than-human world, and assess the role of human agency in both destruction and possible remediation. Mary Gomes (2009) describes the Altars of Extinction project: a series of art installations that act as ritual spaces intended to help viewer-participants "collectively contemplate and grieve the extinction of plant, animal, and fungal species at human hands" (248). The Altars of Extinction project seeks to raise awareness of the current mass extinction spasm and transform ritual participants into activists engaged in the work of preventing future extinctions. The MEMO project (memoproject.org) is a larger project with similar intent.

2 Sigmund Freud (1957a, 1957b; Clewell 2004; von Unwerth 2005; but also see Freud 1960, 386) is generally associated with the detachment theory of grieving. Dennis Klass (Klass 2006; Klass and Goss 1999; Klass, Silverman, and Nickman 1996) has published widely on the continuing bonds theory. A brief overview of theories and models of grief is provided in Wright and Hogan (2008) and theories of grieving are treated more extensively in other contributions to this collection.

3 I acknowledge the students in my first-year seminar at Ithaca College in fall 2013 (Death of Nature: Mourning Environmental Losses) for exploring these materials with me and refining my insights about the relevance of these ritual examples.

4 Details in this paragraph are drawn from Hoffman (1996).

5 The Abrahamic traditions are often overwhelmingly patriarchal. In Judaism, for example, only adult males count in the number required for a minyan. See Rochelle Millen (1990) for an argument for the participation of women in Kaddish practices.

## REFERENCES


Adeyemo, Tokunboh. 1979. *Salvation in African Tradition*. Nairobi: Evangel Publishing House.

Afary, Janet. 2003. "Shi'i Narratives of Karbala and Christian Rites of Penance: Michel Foucault and the Culture of the Iranian Revolution, 1978–1979." *Radical History Review* 86: 7–35.

Alenuma, Sidonia. 2002. "The Dagaare-Speaking Communities of West Africa." *Journal of Dagaare Studies* 2: 1–16.

Ayoub, Mahmoud. 1978. *Redemptive Suffering in Islam: A Study of the Devotional Aspects of Ashura in Twelver Shi'ism*. The Hague: Mouton.

Batt, Herb. 1988. "Sky Burial [Poem]." *Anima* 14, no. 2: 111.

Buddhaghosa, Bhadantacariya. 1976. *The Path of Purification (Visuddhimagga)*. Translated by Bhikkhu Nyanamoli. Vol. 1. Berkeley: Shambhala.

Burton-Christie, Douglas. 2011. "The Gift of Tears: Loss, Mourning, and the Work of Ecological Restoration." *Worldviews* 15, no. 1: 29–46.

Camphausen, Rufus C. "Sky Burial, or Rather Jhator." *Dakini Yogini Central*. http://yoniversum.nl/dakini/jhator.html. Accessed 6 July 2013.

Clewell, Tammy. 2004. "Mourning beyond Melancholia: Freud's Psychoanalysis of Loss." *Journal of the American Psychoanalytic Association* 52, no. 1: 43–67.

Deeb, Lara. 2005. "Living Ashura in Lebanon: Mourning Transformed to Sacrifice." *Comparative Studies of South Asia, Africa, and the Middle East* 25, no. 1: 122–37.

Diamant, Anita. 1999. *Saying Kaddish: How to Comfort the Dying, Bury the Dead, and Mourn as a Jew*. New York: Schocken.

Faison, Seth. 1999. "Lirong Journal: Tibetans, and Vultures, Keep Ancient Burial Rite." *The New York Times*, 3 July. http://www.nytimes.com/1999/07/03/world/lirong-journal-tibetans-and-vultures-keep-ancient-burial-rite.html. Accessed 13 June 2013.

Freud, Ernst L., ed. 1960. *Letters of Sigmund Freud, 1873–1939*. London: The Hogarth Press.

Freud, Sigmund. 1957a. "Mourning and Melancholia." In *The Standard Edition of the Complete Psychological Works of Sigmund Freud, Volume XIV (1914–1916)*, edited by James Strachey, 237–58. London: Hogarth Press.

– 1957b. "On Transience." In *The Standard Edition of the Complete Psychological Works of Sigmund Freud, Volume XIV (1914–1916)*, edited by James Strachey, 303–7. London: Hogarth Press.

Goldenberg, Robert. 1992. "Bound Up in the Bond of Life: Death and Afterlife in the Jewish Tradition." In *Death and Afterlife: Perspectives of World Religions*, edited by Hiroshi Obayashi, 97–108. New York: Greenwood Press.

Goldman, Ari. 2006. *Living a Year of Kaddish: A Memoir*. New York: Schocken.

Gomes, Mary E. 2009. "Altars of Extinction." In *Ecotherapy: Healing with Nature in Mind*, edited by Linda Buzzell and Craig Chalquist, 246–50. San Francisco: Sierra Club Books.

Goody, Jack. 1962. *Death, Property, and the Ancestors: A Study of the Mortuary Customs of the LoDagaa of West Africa*. Stanford: Stanford University Press.

– 1972. *The Myth of the Bagre*. Oxford: Clarendon Press.

Goody, Jack, and S.W.D.K. (Kum) Gandah. 2002. *The Third Bagre: A Myth Revisited*. Durham: Carolina Academic Press.

Goss, Robert E., and Dennis Klass. 1997. "Tibetan Buddhism and the Resolution of Grief: The Bardo-thodol for the Dying and the Grieving." *Death Studies* 21: 377–95.
– 2006. "Buddhisms and Death." In *Death and Religion in a Changing World*, edited by Kathleen Garces-Foley, 69–92. Armonk: M.E. Sharpe.
Grainger, Roger. 1998. *The Social Symbolism of Grief and Mourning*. London and Philadelphia: Jessica Kingsley Publishers.
Heilman, Samuel C. 2001. *When a Jew Dies: The Ethnography of a Bereaved Son*. Berkeley: University of California Press.
Hessel-Robinson, Timothy. 2012. "'The Fish of the Sea Perish': Lamenting Ecological Ruin." *Liturgy* 27, no. 2: 40–8.
Hoffman, Lawrence A. 1996. "Rites of Death and Mourning in Judaism." In *Life Cycles in Jewish and Christian Worship*, edited by Paul F. Bradshaw and Lawrence A. Hoffman, 214–39. Notre Dame: University of Notre Dame Press.
Karma-glin-pa. 2000. *The Tibetan Book of the Dead: or, the After-Death Experiences on the Bardo Plane, According to Lama Kazi Dawa-Samdrup's English Rendering*. Edited by W.Y. Evans-Wentz. Oxford: Oxford University Press.
Kelly, Thomas. 1997. "Opening the Sky Door [Photo Essay]." *Tricycle* 7, no. 1: 80–3.
Klass, Dennis. 2006. "Grief, Religion, and Spirituality." In *Death and Religion in a Changing World*, edited by Kathleen Garces-Foley, 283–304. Armonk: M.E. Sharpe.
Klass, Dennis, and Robert Goss. 1999. "Spiritual Bonds to the Dead in Cross-Cultural and Historical Perspective: Comparative Religion and Modern Grief." *Death Studies* 23, no. 6: 547–67.
Klass, Dennis, Phyllis R. Silverman, and Steven L. Nickman, eds. 1996. *Continuing Bonds: New Understandings of Grief*. Washington: Taylor & Francis.
Konigsberg, Ruth Davis. 2011. *The Truth about Grief: The Myth of Its Five Stages and the New Science of Loss*. New York: Simon & Schuster.
Kübler-Ross, Elisabeth. 1969. *On Death and Dying*. New York: Macmillan.
Kübler-Ross, Elisabeth, and David Kessler. 2005. *On Grief and Grieving: Finding the Meaning of Grief through the Five Stages of Loss*. New York: Scribner.
Kuukure, Edward. 1985. *The Destiny of Man: Dagaare Beliefs in Dialogue with Christian Eschatology*. Frankfurt am Main: Peter Lang.
Lamb, Robert. 2011. "How Sky Burial Works." http://people.howstuffworks.com/culture-traditions/cultural-traditions/sky-burial.htm. Accessed 13 June 2013.
Lamm, Maurice. 1969. *The Jewish Way in Death and Mourning*. Middle Village: Jonathan David Publishers.
Leopold, Aldo. 1949. *A Sand County Almanac and Sketches Here and There*. New York: Oxford University Press.

Levin, Dana. 2011. "Cathartes Aura [Poem]." In *Sky Burial*, 7–10. Port Townsend: Copper Canyon Press.
McKibben, Bill. 1989. *The End of Nature*. New York: Random House.
Millen, Rochelle L. 1990. "Women and Kaddish: Reflections on Responsa." *Modern Judaism* 10, no. 2: 191–203.
Nakash, Yitzhak. 1993. "An Attempt to Trace the Origin of the Rituals of Ashura." *Die Welt des Islams* 33, no. 2: 161–81.
Norton, Augustus Richard. 2005. "Ritual, Blood, and Shiite Identity: Ashura in Nabatiyya, Lebanon." *TDR (1988–)* 49, no. 4: 140–55.
Sambhava, Padma. 1993. *Tibetan Book of the Dead: The Great Book of Natural Liberation through Understanding in the Between*. Translated by Robert A. Thurman. Westminster: Bantam.
Shaw, Sarah. 2009. *Introduction to Buddhist Meditation*. New York: Routledge.
Sindawi, Khalid. 2001. "Ashura Day and Yom Kippur." *Ancient Near Eastern Studies* 38: 200–14.
Somé, Malidoma Patrice. 1993. *Ritual: Power, Healing, and Community*. New York: Penguin.
– 1994. *Of Water and the Spirit: Ritual, Magic, and Initiation in the Life of an African Shaman*. New York: Putnam.
– 1998. *The Healing Wisdom of Africa: Finding Life Purpose through Nature, Ritual, and Community*. New York: Putnam.
Thubron, Colin. 2011. *To a Mountain in Tibet*. New York: HarperCollins.
*The Tibetan Book of the Dead*. Videorecording. Directed by Hiroaki Mori, Yukari Hayashi, and Barrie McLean. 1994. New York: Lorber HT Digital, 2009. DVD, 90 min.
Turner, Victor W. 1969. *The Ritual Process: Structure and Anti-Structure*. Chicago: Aldine.
van Gennep, Arnold. 1960. *The Rites of Passage*. Translated by Monika B. Vizedom and Gabrielle L. Caffee. Chicago: The University of Chicago Press.
von Unwerth, Matthew. 2005. *Freud's Requiem: Mourning, Memory, and the Invisible History of a Summer Walk*. New York: Riverhead Books.
Wieseltier, Leon. 1998. *Kaddish*. New York: Vintage.
Windle, Phyllis. 1992. "The Ecology of Grief." *BioScience* 42, no. 5: 363–6.
Wright, Patricia Moyle, and Nancy S. Hogan. 2008. "Grief Theories and Models: Applications to Hospice Nursing Practice." *Journal of Hospice and Palliative Nursing* 10, no. 6: 350–6.

# 3 Mourning Ourselves and/as Our Relatives

*Environment as Kinship*

SEBASTIAN F. BRAUN

Grief over losses in the natural world has become a common part of our personal and collective emotional landscape. At the same time, the very depth and extent of these losses has produced, for many, a kind of psychic numbing, resulting in an inability or unwillingness to acknowledge or respond to this loss lest it completely overwhelm and perhaps debilitate us. There is too much to grieve … The diminishment or loss of the capacity to mourn reflects a loss of knowledge, an unraveling of the ties of kinship that bind us to the lives of other beings.

DOUGLAS BURTON-CHRISTIE

## Introduction

I want to begin this exploration with three statements about mourning and the environment. The first two are related to the longstanding discussion about buffalo in Yellowstone National Park. Because they are suspected of being exposed to brucellosis, the state of Montana has engaged in a variety of plans and strategies to kill these buffalo upon entering its territory. Activists, both Native and non-Native, have long protested these practices. As part of their argument for better treatment of the buffalo, they have been portraying the buffalo in ways that show similarities in these animals to human behaviour. One buffalo behaviour that is often mentioned in this regard is mourning: "Tell FWP [Montana Fish, Wildlife, and Parks] that you expect them to better educate hunters on the behavior of wild buffalo. Buffalo will mourn their fallen brothers and sisters, and if a hunter insists on killing a buffalo, they must respect them enough to give them the opportunity to do this"

(Buffalo Field Campaign 2006). Buffalo mourn each other. At least that is the meaning that some people attach to the fact that when a buffalo is shot, others will gather around. The meaning humans attach to this social activity is "mourning." It is not simply the case that buffalo deserve to be able to mourn for each other, however. People mourn buffalo too: "We urge people who care about wild buffalo and wild lands to speak out in powerful silence to stop the slaughter, while we mourn the loss of those already gone" (CampusActivism.org 2004). In 2004, the Buffalo Field Campaign organized an event "to mourn the senseless deaths of nearly 300 [buffalo] killed this year alone." When I worked with a tribal buffalo operation on the Cheyenne River Sioux Reservation, we always had a ceremony after a buffalo was killed.

Some Lakota mourn not only individual buffalo, but also the species in its splendor, in freedom, in its natural environment, and as a symbol of historical trauma. Through the buffalo they also mourn the time of Lakota traditional culture and its dominance over the northern plains, since buffalo and historic Lakota culture are seen to be in many ways intimately connected (Braun 2008). Many people also mourn their pets, cows, wolves, sheep, and, as the third statement shows, a variety of other losses in our daily lives: "'Grass no good upside down,' said a Pawnee chief in northeast Colorado as he watched the late-nineteenth-century homesteaders rip through the shortgrass with their steel plows. He mourned a stretch of land where the Indians had hunted buffalo for millennia" (Popper and Popper 1987, 16). I am less interested, here, in the event or emotion of mourning for environmental loss. Instead I would like to take a look at what the fact that people do mourn such loss tells us about human-environmental relationships.[1] Implicit in the practice of mourning, I argue, is that we relate to our environment through kinship relations. Mourning is a social activity, extended to close friends and relatives. My contention, then, is that people mourn for specific parts of their environments because they feel related to them. Judith Butler wrote that, in mourning, "something about who we are is revealed, something that delineates the ties we have to others, that shows us that these ties constitute what we are" (2004, 22). Mourning for the environment, in other words, shows what kinds of kinship relations we create and lose with our environments.

## Environment and Kinship

The assertion that people mourn for animals, landscapes, or other parts of the environment might be relegated to the realm of anthropological truth, or perhaps to the equally "exotic" categories and practices of totemism or animism, where such relations are common. For example, the authors of a study on *Indigenous Kinship with the Natural World in New South Wales* assert that "the connections between people and parts of the natural world in bonds of mutual life-giving was impressed upon us by all of the people with whom we spoke" (Rose, James, and Watson 2003, 2). Although totemistic kinship analysis has reduced "relationships of mutuality to the strictly human side," as Rose et al. (2003) point out, their consultants "consistently affirmed the mutuality of these bonds," and constructed "a worldview, in which humanity is part of the natural world, has responsibilities to the world, and is born from, lives for, and dies to return to, the living world known as country" (2).

This worldview reflects how kinship connections make relatives of animals, plants, landscapes, spiritual beings, and other non-human entities. Because American culture is based on a conceptual difference between human and natural realms, and because it sees kinship as biological (Schneider 1980), in mainstream culture these non-human entities are not categorized as relatives, but as parts of our religious or biological environment. Kinship with such non-human beings is, therefore, often perceived as metaphorical, a mere cultural invention, a symbolic relation at best. I hasten to emphasize that they are as real as any other kinship relation. In the constitution of relationships of rights and obligations between relatives, and in the consequential fulfillment of these relations, they are not merely metaphorical. McKinley (2001) rightly asserts that "such a view seriously underestimates both the power and scope of the philosophy of kinship. The supernatural kin are no more metaphoric than the human ones. They may be 'more intangible,' but they are not more metaphoric. All kinship involves metaphor right from the start or it could not have any moral and ethical content at all" (132). A relegation of kinship relations with non-human entities to the anthropological realm can create the idea that totemistic kinship relations carry no real consequences for non-Indigenous, and therefore, as is still often assumed, rational and scientific societies. In this way, so-called

Western societies are often portrayed as free from spiritual, metaphorical illusions, either because they know that religion is simply an opiate for the people, or because their religion – which supposedly teaches the superiority of humans over other beings – is the true religion, and other worldviews are simply caught in primitive, magical, myths.

However, as French anthropologist Claude Lévi-Strauss explained in *The Elementary Structures of Kinship*, both of these arguments are wrong. The latter argument, which he called the "archaic illusion" (1969, 84–97), is simply an exercise in ethnocentrism, while the former rests on an impossible possibility of stepping outside cultural relationships. This, however, is impossible, since "the species has no natural behaviour to which an isolated individual might retrogress" (5) – nor, I should add, to which a society can ascend by leaving culture or kinship behind. Whether or not a society is kinship-based is only a categorization of their political organization; it cannot mean that some societies recognize kinship and others do not. While values of extended family, friendship, and community have decreased in industrialized societies, nuclear family ties have gained more importance (Fowlkes 1990). Although the extension of kinship networks, and thus the applicability of kinship to everyday lives and encounters, has been diminished in some societies, there are no societies without kinship.

Different cultures construct kinship in different ways, of course. Nonetheless, relationships with non-human entities form the basis for any relation to the environment, even in American society. Such relations, it is often argued, are based on ethical or moral foundations. For example Gregory Cooper has made the argument that Aldo Leopold's *Sand County Almanac* opened up a new ethic, a "project for moral theory;" it does so, Cooper argues (1996, 151), because Leopold calls for the recognition of biotic communities, and "morality is the glue that binds human communities." Indeed, Leopold argued, "the practice of conservation must spring from a conviction of what is ethically right, as well as what is economically expedient" (1991, 345). Langdon Gilkey writes that the "political decisions that we make for or against nature depend on our most spiritual attitudes, namely, our moral obligations and our moral decisions" (1993, 149). It is the term "obligation," however, that leads me to a slightly different emphasis. Leopold writes that "obligations have no meaning without conscience, and the problem we face is the extension of the social conscience from people to land" (1991, 341). Extending social conscience, and the obligations that go with it, from people to land means to extend kinship obligations to the environment. Gilkey recognizes this when he defines human relationship to nature as "our sense of kinship with nature; correspondingly, our sense of respect and of moral obligation to nature" (1993, 149). Moral obligations exist as concrete kinship obligations.

This is important because a retreat to morality as the only basis for environmental relations leaves the door open for individualistic interpretations of moral values, for a denial or superficial acknowledgement of community and the obligations, responsibilities, and rights that a real community entails. To deny any kinship relationship with – that is, any social obligation to and responsibility for – non-human entities is truly to see the world as "no more than a notional constraint against which the ego may flex its muscles and delight in its powers, a convenient springboard by which it can recoil onto itself" (Eagleton 1990, 123). It is kinship that creates moral obligations and a social conscience that creates a community. It is, therefore, kinship relations that are expressed in mourning for the losses of community members. These relations might be perceived, felt, and expressed in paradoxical ways but as Julie Cruikshank (2004) notes, "social relations are rarely straightforward" (27).

As there are no societies without kinship, even the most urbane societies, even people who might live almost completely alienated or solitary lives, still make relatives and attach important meaning to kinship practices. The fact that these are extended to the non-human environment can be seen in the example of pets. Henry Sharp (2001, 86–7) maintains that, because dogs are seen as persons within the family, dogs provide "the most coherent definition of family in American culture." He maintains that the distinction between the dog as person (in the family) and as pet (outside the family) "is the clearest marker of family in American culture." Indeed if we mourn more for our dog (a relative, and therefore a person) than for the neighbour down the street (a stranger, and therefore non-person) it is a clear indication that the dog has closer kinship ties to us than does the neighbour. Whatever else that says about a society, it is an indicator that people can and do create kinship ties with their environments. Marshall Sahlins (1976) pointed out almost four decades ago that American culture treats dogs and horses differently from pigs and cows (or "cattle") – the latter are eaten – and that these relationships are partly visible through naming practices, that is, through the establishment of intimate kinship ties. The difference in relationships to cattle – cows raised for meat – versus cows – dairy cows, or sometimes, as in eastern Africa, also cows used for their blood – also follows this pattern. Many times cows have names, and it is not unusual that people talk with them; most cattle are anonymous. They are excluded from kinship because they will be slaughtered.

Kinship does not actually prevent violence, though: some animals slaughtered by small-scale farmers have names, and their owners talk with them. Extension of personal relationship does not save them from serving as a food source. There is a difference, however, between being slaughtered anonymously and being slaughtered by kin. One is akin to massacre (or battle), the other akin to ritual (or sacrifice). While the former implies the absence of any obligations, the latter comes with specific responsibilities. The fact that kinship ties do not necessarily prevent humans from killing animals – or, in general, from causing environmental loss – indicates clearly that not all relationships are the same. As Michael Jackson (2013) notes, "violence is not an expression of animal or pathological forces that lie 'outside' our humanity; it is an aspect of our humanity itself. Rather than dismiss it as antisocial behavior ... we must approach it as a social phenomenon whose conditions of possibility inhere in the 'three obligations of reciprocity – giving, receiving, repaying'" (171). For example, historically, Lakota people killed buffalo to survive. Buffalo hunting was, in part, an expression of their kinship relation to buffalo. A romantic perspective on human-environmental kinship ties can lead us to believe that kinship means harmonious coexistence. But as Karen Warren (2000) points out, "all living things" were not "brothers or sisters, uncles or mothers, grandfathers or other members of a family" (86–7). Such a characterization of universal relations, in terms of close kinship with the implicatons of human social values and reciprocity, is misleading and simplifying, not least because it does not translate different cultural notions of familial relationships. John Grim argued that many "Indigenous peoples generally regard species as unique in their own particularity, but not necessarily equal" (2001,

51). All cultures are selective about whom to include in what kinds of kinship relations, and about what meanings to assign to what specific, selected parts of their environments.

At the most extreme, sometimes killing another being can actually create kinship, perhaps because the anonymous killing of another (often human) being would be too dehumanizing for all involved. This was the case, for example, with some historic headhunting cultures. In some New Guinean societies, the primary name of men was the one acquired from the victim of such violence (e.g. van der Kroef 1952, 222). Similarly, it is the fighting and overpowering of spiritual beings that results in the establishment of a relationship with them and thus leads to the gain of knowledge/power for the Mission Dene, Sharp (2001, 111) notes. David Anderson describes this for the Evenki in Siberia, whose attitude toward hunting success is "an entitlement through knowing the land" (2000, 129). The relevant social environment here is "a social community of kinsmen and animals," in which both come to know the other (2000, 124–30). Kinship relations, then, do not only involve the acceptance of self-sacrificial offering, but sometimes the violent taking for our own interest. Kinship relations do not imply equality or harmony. Establishing and re-enacting dominance can be a necessity for kinship relations. For the Lakota, for example, the fact that the two-legged won a race over the four-legged is often cited as the legitimization for humans killing buffalo. These relationships, in order to be kinship relations, also impose obligations. One of them might very well be mourning.

## Beyond the Exotic

In discussions of environmental relations, it seems tempting to see the practice of mourning and its absence through a romantic lens. This is true especially in the context of Indigenous peoples, who are often defined as having a special relationship with nature – richer, deeper, closer, and more spiritual than so-called Western societies (e.g. Niezen 2009). This also fits with a whole range of assumptions starting with the one that Indigenous societies are kinship-based, while Western ones are not. Raymond DeMallie (1998) has explained that in "virtually every American Indian society kinship is culturally defined to include the relationship of human beings to all other forms of existence in a vast web of cosmic interrelationships ... As a cultural phenomenon it defines relationships (sets of statuses), prescribes normative patterns for behavior among relatives (roles), and extends those patterns outward to the universe" (306). In light of relationships to non-human entities, it might be tempting to see these kinship systems "in striking contrast to the static nature of American kinship, an ascribed system of roles whose behavioral content tends always to be minimized" (DeMallie 1994, 132). This contrasting view expressed by DeMallie's collaborators, however, might be more informed by hegemonic "assumptions and presuppositions" about contrasting cultures and therefore kinship systems (Schneider 1984, 196) – that is, this view is more supported by identity politics than by ethnographic facts. Thomas Thornton asserts this in the context of relationship to places, but his statement can easily be extended to relations with other parts of the environment: "Most of the popular literature on indigenous peoples' conceptions of places has tended to contrast implicitly or explicitly 'their' intimate, enchanted union with the landscape with 'our' (Euro-American) mechanistic or estranged view. Obviously, such an approach is limited, if not flawed, and often yields a superficial or one-dimensional view of the complex and unique relations of peoples to the places they inhabit throughout the world" (2008, 7). To deny the romantic notion of the "noble savage" opposite the modern, technologically driven Western destroyer of the earth is to deny the shedding of responsibility. Assuming that industrial societies innately lack connection to the environment gives them *carte blanche*. To remind people that all societies create kinship ties with their environments, however selective, and that all societies therefore incur obligations to their environment (including mourning practices) means the reestablishment of the notion of responsibility. The capacity to mourn for environmental losses is not an emotional response, born of a culturally different morality that "we" no longer hold on to. It is an expression of culturally defined kinship obligations. As the complex layers of mourning relationships are unravelled, it becomes clear that mourning for environmental loss is neither restricted to historical societies nor to Indigenous cultures. This opens up at least two issues: the notion of non-anthropocentric ecology and the notion of fundamental difference.

Ecology as kinship simultaneously denies the possibility of non-anthropocentric environmental relations and makes clear that in these relations, humans do have obligations to non-human relatives. What Grim (2001) writes about deep ecology, namely that it "can be distinguished initially from a particular ... tradition's environmental knowledge as oriented more toward individuals than communities" (39), applies to non-anthropocentric philosophies. They, too, approach "the world as a call to a way of knowing that has philosophical, moral, and political implications *for individuals*" (ibid., emphasis added). Paradoxically, non-anthropocentric views are individualistic philosophies. They are outside the realm of community practices because meaning is constituted by community (Sharp 2001). Rana and Marcellini (2012) write that "objects, all the way from lightbulbs, clouds, and *Two and a Half Men* to salamanders and the memory of a first kiss, are just as reality producing as we are. All these things must be understood as social objects, endowed with agencies and allusive powers that we can only apprehend sympathetically, at a distance." If a non-anthropocentric reality exists, however, it carries no meaning for humans, and thus remains culturally and socially irrelevant. We can imagine such a reality, but we cannot relate to it: to relate to anything is to give it meaning, and meaning is always cultural, that is (for humans) anthropocentric.

It is the necessity of cultural, community-shared meaning for relationships that raises the issue of difference. Val Plumwood makes clear that the assumption that we know what meanings other beings – communities or animals – attach to reality can lead to what she calls "oppressive projects of unity" (2000, 66–9). While Eric Katz writes that "I may identify with the deer sitting in the bushes outside my window, in that I recognize our common interests in the maintenance of this habitat" (2000, 26), Plumwood questions the recognition of a common interest, and sees it more as an imposition. "We may identify in solidarity with an animal," she notes (2000, 63), "but we do not thereby acquire identical specific interests ... We must attain solidarity with the other *in their difference*." It is this difference, and the recognition and acknowledgement of difference, which allows us to get away from the exotic, romantic imagination of the environment and of other peoples' cultural practices. Kinship is not necessarily solidarity in difference, but it is definitely relationship – mutual obligation and rights – in difference. We are mourning not ourselves, but others. The fact that we mourn, however, is an expression of relation(s).

## Mourning

John Bennett posed the question "whether or not the *social* environment – humans interacting with humans and their creations – is really a legitimate part of an ecological approach," that is whether "the 'environment' of human society has anything in common with the 'environment' of Nature" (1996, 365). I argue that kinship relations, in part shown through mourning practices, demonstrate the affirmative answer to both questions: "where animals and humans share common states of being that include family relationships, intelligence, and mutual responsibility for maintenance of a shared world, interaction with the physical world is a social relationship" (Cruikshank 2004, 27). The second question raises another fundamental issue: does "mourning" for "the environment" carry the same meaning cross-culturally?

Burton-Christie implies that mourning and grief for the environment is a practice that goes back to pre-industrial cultures, and therefore that mainstream society's perceived incapability to mourn for the environment reflects "a loss of knowledge, an unraveling of the ties of kinship" (2011, 30). This idea ties in with the idea of Indigenous societies – purportedly defined as non-industrial – mourning for their industrially spoiled environments. However, in the case of the anonymous Pawnee chief in Popper and Popper's account quoted above, the idea of mourning is the Poppers', not necessarily the chief's. If mourning for environmental loss is a response to industrial destruction, non-industrial cultures do not need it; the concept might have been imposed on practices that carry a different meaning in their cultures, just as we cannot be sure if buffalo really mourn one of their own, it just seems that way from our own cultural interpretation of their behaviour. Could it be that the idea of mourning for the environment is an invention of industrial societies, an idea that originated with the romantic, aesthetic response to industrialization, and that non-industrial societies have no use for it? If so, mourning for environmental loss, as defined and imposed on others by American values, might be an expression of nostalgia, qualitatively different from mourning.

Perhaps the best known anthropological example of a ceremony held for dead animals is Hallowell's bear ceremonialism (1926), and other related rituals. In his description of the ceremonial practice, Hallowell actually uses the term mourning albeit in quotation marks. In some areas, he writes (59), "we find that a 'mourning-song' is chanted after the animal's death," adding that its "function seems to be analogous to the conciliatory speeches of other peoples." It is clear that the bear is considered a relation, but the ritual seems to revolve much more around the ability to hunt other bears in the future, that is the continuation of the relationship with the entity "bear," than around mourning for the individual (see also Barbeau 1945). Hunting in many northern societies, writes Ingold (2000), "is conceived as a rite of regeneration: consumption follows killing as birth follows intercourse, and both acts are integral to the reproductive cycles, respectively, of animals and humans. However, animals will not return to hunters who have treated them badly in the past" (67). Most traditional ritual practices around real, feared, or imagined environmental loss might indeed be restorative rituals, ensuring that the overarching relation to the meaningful parts of the environment will continue (see also Nesper 2002, 36–41). This is true for specific hunting ceremonialism as well as the Sundance, and for green corn ceremonies as well as blessings of the fields or praying against droughts.

To mourn is always an expression and fulfillment of kinship. The capacity to mourn for relatives stands at the centre of relationship making because it retroactively validates kinship relations. Burton-Christie writes that the "ability to mourn for the loss of other species is, in this sense, an expression of our sense of participation in and responsibility for the whole fabric of life of which we are a part. Understood in this way, grief and mourning can be seen not simply as an expression of private and personal loss, but as part of a restorative spiritual practice that can rekindle an awareness of the bonds that connect all life-forms to one another and to the larger ecological whole" (2011, 30).

Mourning as an act to express participation and responsibility is a public performance, a public acknowledgement, a ritual practice. As a public performance of social obligation, it matters not whether the participants actually grieve. What matters is the participation in the ritual, which reaffirms the obligations resulting from the relationship (Rappaport 1999, 119–22). These obligations go beyond the death of a particular relative, and instead pertain to the relationship of two whole social entities.

If, as seems to be true for bears in the north, the general relationship with an entity regulates the reappearance of individual manifestations of this entity (see Sharp 2001; Berkes 1999, 79–93; Krech 1999; Mills and Slobodin 1994), then mourning for an individual is not the primary obligation; what needs to happen instead is assuring that other individuals will return. This requires power, and power is tied to knowledge (Sharp 2001; Anderson 2000, 127–31). Larry Nesper (2002) shows that in Ojibwe thought "human beings need spiritual power gained from relationships with non-human persons in order to realize *pimadaziwin*, life in a complete and full sense" (34). Howard Harrod argued that appropriate relations with non-human powers were required in order to be human on the northern plains, and that being human "was fundamentally intertwined with a relation to particular places and specific animals" (2000, 43). Both Nesper and Sharp, however, describe a system in which relationships to specific individual animals are much less important than relationships to the entities that control/represent these animals. David Anderson (2004) writes that for Evenki and Inuit, "there is a notable absence of the hypothesis that life is to be enjoyed by individual creatures. Instead it is thought to be something that neighbouring organisms share with each other" (14). Mourning as a regenerative, productive ritual ensures that relationships are transferred from those dead to those living. In hunting communities, it might be a reminder and a plea for animal spirits that the killing of an individual animal did not eradicate the fundamental relations. In all cases, it is a statement on the fact that relationships, and thus obligations and responsibilities, still exist: life goes on.

Indigenous peoples are sometimes invoked in mourning for a "geography of loss" (O'Nell 1996, 117). When they are, campaigns often play on two popular notions of non-industrial peoples, "one linking hunters with harmony, the other conflating norms with behaviour" (Cruikshank 2004; 27). In reality, however, Indigenous peoples live and lived in complex situations that often do and did not leave room for mourning environmental loss in a romantic or aesthetic fashion that would lead to preservation for preservation's sake. Aestheticism or romanticism are views from afar, luxuries afforded to nostalgia and to people who have become alienated from daily interactions with their environments and no longer recognize that life includes loss. This is not to say that people do not experience or appreciate loss. As Fikret Berkes (1999) notes, "the hunter loves the animals he kills" (91). However, to stop inflicting everyday environmental loss – losses from hunting, farming, or other normal cultural activities – is often not an option, and might lead to larger negative consequences because it would alienate pieces of the environmental whole. As Wishart (2004) notes, "'wild' is not considered to be positive because it signals the breakdown in the human-animal relationships. 'Wilderness' is a horrific concept for Gwich'in because it refers to situations where people can no longer enter into relationships with animals and starvation and social breakdown are imminent" (86). Romantic efforts to preserve life for preservation's sake thus threaten some societies as much as abnormal, outside, large-scale, uncontrollable environmental destruction. They both represent the loss of normal relations that fundamentally threaten the people.

Similarly, the feelings of loss, grief, and mourning that Cunsolo describes for her Inuit colleagues (chapter 7) are emotions over the loss of fundamental relations that endanger the very survival of the people as such. The anticipatory grieving she describes, then, might be more than a grieving for environmental loss. It might also be an anticipatory grieving for the loss of the culture or the people.

This grieving over climate change and its related impacts to environment is also, of course, a reminder that, as Anderson notes, Circumpolar societies, just like Siberian reindeer herders, are "as firmly rooted in the dilemmas produced by industrial development ... as are the lives of any factory-worker in Yekaterinburg or Manchester" (2000, vi). The global threat to fundamental relations, and the mourning of that kind of environmental loss, whether anticipatory or reactive, is new, but this is not to say that all mourning for environmental loss is only a recent, industrial practice. My point, on the contrary, is that non-industrial peoples do not differ fundamentally from industrial societies (Feit 1994): kinship with the environment is neither a recent invention nor should we expect it to have disappeared. Nevertheless with global industrialization, a new quality of loss, and therefore of mourning, has been introduced. At the same time, the loss of real kinship relations through alienation has made it possible or perhaps necessary to imagine such relations, to practise these relations, and to ascribe these same practices and the emotions surrounding them on others.

## Alienation

Alienation is often seen through a socio-economic lens. Industrialization leads to the alienation of the worker from the product as well as the increasing alienation of people from each other. Thus, Fowlkes writes, "the combined forces of industrialization and urbanization have substantially eroded the value assigned to attachments based on community, extended family ties, and friendships. At the same time that these ties have diminished in importance, the meaning of nuclear family relationships has intensified" (1990, 638). If relations to the environment are kinship-based, however, alienation must have ecological effects too. John Bellamy Foster (2000) argues that the ecological consequences of alienation result, on the one hand, from the effects of capitalism on agriculture as the yeoman farmer, the peasant with roots in rural communities, disappeared. On the other hand, Foster notes, "alienation from the earth is the *sine qua non* of the capitalist system" (174). The relationship between alienation from the earth, the transformation of agricultural communities from peasants to wage labourers, and the loss of community has been decried by Wendell Berry. He argues that there cannot be "a postagricultural world that is not also postdemocratic, postreligious, postnatural – in other words, it will be posthuman, contrary to the best that we have meant by 'humanity'" (1996, 78). In this sense, alienation, because it severs community and kinship ties, will indeed lead to a "diminishment or loss of the capacity to mourn" (Burton-Christie 2011, 30).

Fredric Jameson (1998) enlarges this argument and applies it to relations with landscapes. Capitalism drives parcellization, the commodification of space, and traditional cultures follow peasant agriculture into extinction, he notes (65–8; see also Buckland 2004). The ultimate consequence of this process, however, goes further: "Where the world system today tends toward one enormous urban system ... the very conception of the city itself and the classically urban loses its significance and no longer seems to offer any precisely delimited objects of study, any specifically differentiated realities. Rather, the urban becomes the social in general, and both of them constitute and lose themselves in a global that is not really their opposite either" (Jameson 1998, 69). The people who live in rural areas do not live any differently from people in urban areas

(Danbom 2006, 258–63). In fact, increasingly fewer interactions with the environment mean that it does not matter where one lives. McMansions can be built anywhere – suburbs and exurbs look alike wherever they are – and posters or screens could replace panoramic views. "For most North Americans, the modern home, whether a single-family suburban residence or a multistory urban apartment building, has been fundamentally 'de-physicalized,' or disconnected from the landscapes and ecosystems that support it" (Beatley 2004, 210). Increasing numbers of people are thus deterritorialized and live in what Marc Augé (1995) has called the "non-places" of supermodernity and David Ehrenfeld (1996) calls "pseudocommunities." Alienation, from Ehrenfeld's (1996) perspective, has reached the point at which people live in a hybrid world, in which the real world "is confusingly intermingled with the artificial world of electronic communication." In these pseudocommunities, "many of the human threads and connections that once bound people together into working communities" have been cut (20, 21).

In the non-places "of supermodernity people are always and never at home;" the "space of non-place creates neither singular identity nor relations; only solitude, and similitude" (Augé 1995, 103, 109). This cannot be surprising. As Thornton notes, the "experience of making a living in an environment is central to what it means to 'inhabit' a place as opposed to simply observing or passing through it" (2008, 25). It seems that passing through the environment, not rooting in it, defines how many North Americans and, increasingly, other people live globally. Beatley (2004, 18) thinks the transient lifestyle, which prevents us from developing knowledge of places, is a modern phenomenon. Jackson (1994), on the other hand, thinks that it was the temporary nature of settlements that produced the "landscape of empty, interchangeable divisions" (154–5) as far back as the early nineteenth century. Regardless, it is obvious that non-places and non-relations deny the possibility to build relationships with landscapes and environments whose losses we could mourn. For Indigenous as well as for immigrant and diasporic peoples, Thornton (2008, 190–1) points out, "a healthy sense of place requires participatory engagement in the plots, characters, and settings that compromise ancestral landscapes as cultural-ecological systems." It is not impossible to build places of cultural significance. People can become "native" to a place. But in order to do so they have to become – to quote Basso (1996,

7) – "the place-worlds [they] imagine." In order to feel loss, to be able to mourn, we need to become related.

It would be very tempting to conclude, then, that alienated societies simply do not mourn for the environment or parts of it: that indeed these societies lose the capacity to mourn. However, that conclusion would be wrong. In some ways, living in non-places, the alienation from the specific environment, from specific relatives living within the same communities, has opened the possibility to feel related to anything. If it does not matter where we live, if we are always and never home, if the specificity of relations has been washed away by transience, we can attach meaning to anybody, anything, anywhere. Ehrenfeld's pseudocommunities are not non-communities. They are simply communities that do not exist in the real, geographical, ecological sense of the term. As Fowlkes (1990) notes: "ours is the kind of society that provides both opportunities and needs for intimacy that frequently extend far beyond what is available from our families of origin especially, but quite often from our families of marriage as well. Yet no matter how many or how meaningful our extrafamilial attachments to others, they are privatized, individually negotiated, and lack the moral imprimatur of the love normatively associated with formalized family ties" (638). When relations are individualized, loss and grief are also private and individual, lacking a ritual acknowledgement and expression. Mourning stops being a cultural practice and becomes a psychological phenomenon. However, as Fowlkes (1990, 638) argues, it "is not the category of relationship loss that occasions grief but, rather, the loss of significant attachment." Grief does not become less because of alienation; it simply changes its quality. One consequence of living in a relatively alienated society, in which kinship relations have become individualized, is that "we" do not mourn for "them" anymore, instead "I" mourn for "him/her/it." Pseudocommunities replace the community. One consequence of living in privatized non-places, yet still as humans with opportunities and needs for meaningful relations, is that people mourn those with whom they are related through pseudocommunities. Benedict Anderson's imagined community (1991) – the idea that unrelated people have to imagine themselves as a community to provide stability to large political units – has transformed itself from the state to imagined privatized networks. People mourn for Lady Diana, for Prince, and for victims of crimes globally whom they have never known. Thousands of others, dying of the same causes, but with whom we do not imagine a relation, can be ignored.

Ideas about collective, cultural mourning shift with the expansion of the modern state, advocating and enforcing the idea of citizenship and individual rights, and the concurrent idea of nuclear families as relevant units of kinship. Anderson (2004) thinks that "hierarchically organised state structures of any stripe seem to intensify [the separation between people and animals] rather than alleviate it" (13). They might do so, in part, by denying a "philosophical framework where animals and humans share common states of being" (Cruikshank 2004, 27). The shift in obligations from a focus on individuals as representations of the collective to individuals in and of themselves is simply an indicator of larger changes in the construction of social relations and cultural values that go along with them. For people rooted in a social relationship with their environment, mourning encapsulates relationships between the groups of which individuals are unique manifestations. It is that aspect of mourning that might arguably be more important for cultural environmental relations.

Our relations to our environment demonstrate our kinship behaviour and values. If we do not want to acknowledge that, it might be because the image in the mirror is not one we like to see. Alienation has provided a global expansion of individualistic mourning, but at the same time a fundamental limitation to public rituals of mourning, and thus a limitation to whom we regard as relatives. It has reshaped the boundaries between persons and objects, but not erased the structure of how we relate. Sahlins (1976) wrote that he wondered whether totemism "has not been replaced by species and varieties of manufactured objects, which like totemic categories have the power of making even the demarcation of their individual owners a procedure of social classification" (176). Absolutely alienated, individualistic norms mark the absence of a functional, emplaced community, and because humans are social and cultural animals, this potential social reality usually goes against cultural values. This is still true for a consumer society because as Sahlins (1976) points out, consumption is exchange (177). However, social relations become less important the more important markets become. Once the participants in the exchange do not have to be known, anymore, such a purely economic market "cannot take into account values like the quality of human interaction, culture, or the desire for a healthy environment" (Jaffee 2007, 19). As kinship obligations have been replaced by commodified exchange and aesthetics as the basis for relations, relationships to the environment have changed. Augé (1995) thinks that the "relationship with history that haunts our landscapes is being aestheticized, and at the same time desocialized and artificialized" (73). But he writes about all different types of relationships with the environment.

## Privately Selective Mourning

David Orr (2003, 181) points out that "we are witness to death on the largest scale imaginable – that of life on earth itself." Yet, he argues, "no culture has ever taken greater pains to deny mortality." Refusing to be affected by deaths is also denying relationships that would lead us to mourn or take restorative action. Judith Butler (2004, 32) thinks that when we accept deaths caused by us or in our names "with a shrug or with self-righteousness or with clear vindictiveness," this reaction functions as a "defense against the apprehension of loss." I see clear structural similarities between her argument on the results of military actions and the way we deal with our environments.[2] "When grieving is something to be feared," writes Butler (2004, 29–30), "our fears can give rise to the impulse to resolve it quickly, to banish it in the name of an action invested with the power to restore the loss or return the world to a former order, or to reinvigorate a fantasy that the world formerly was orderly." One of the ways we can adjust to a world so full of loss that grieving becomes impossible is that we can limit who is a relative and what is grievable. If we limit the grievable to those closest to us, death and destruction of others can be accepted as normal, to be shrugged off. Public grieving is still possible, but the "public will be created on the condition that certain images do not appear in the media, certain names of the dead are not utterable, certain losses are not avowed as losses, and violence is derealized and diffused" (Butler 2004, 37–8). Denial of death, or of death affecting us, is the absolution that lets us carry on without having to think about relations and responsibilities.

Relationships have to be selective if we are overwhelmed by quantity. "While we make few landscapes, while we are seldom insiders, we are daily bombarded with hundreds of images of other landscapes, real and imaginary," notes Riley (1997, 209). It is not just landscapes that bombard us, but images and imaginations of ecosystems, climates, animals, plants, rivers, locations, places, and more. With some of these environments or parts of environments, we can relate. We do not have to travel there, see, touch, feel, taste, or run away from them to do this. In our privatized, individual global pseudocommunities, to activate kinship we just have to feel related. Meaningful relations do not require real, sensory interactions, reciprocity is not based on real exchange, and kinship becomes devoid of meaningful behaviour; in post-cultural kinship, a private, emotional connection is enough. This long-distance or imagined kinship is sufficient to trigger mourning for the environment. I do not have to have ever seen polar bears, swum with whales, hiked into reindeer calving grounds, touched coral reefs, or breathed in high-altitude glacier air to feel anticipatory mourning. However, because the relationship is individualized and privatized, this mourning – unless a sufficient number of private individuals mourn the same entity – is not actualized in practice, apart from private acts, and remains a psychological emotion. I can thus mourn polar bears but continue to drive a Hummer.

The paradox of pseudocommunities is exactly that relations can be based on ignorance of each other – the direct opposite of kinship relations, which presuppose knowledge. Kinship is replaced by the superficially aesthetic or, worse, by deep commodification. As Todorov (1984) showed, knowledge "does not imply love, nor the converse; and neither of the two implies, nor is implied by, identification with the other" (185). As Niezen describes, "cultural boundaries are blurred in the unrepressed Web literature by the absence of limits on representation. Uncensored cultural representation makes possible the presentation of community ideals that originate in no recognizable community. More than ever before, it has become possible to express nostalgia for times one has never experienced and pride towards peoples among whom one has never belonged" (2009, 58). These new opportunities to imagine community also apply to environmental relationships. While relations to local environments are intensely privatized, the mobilization of anticipatory mourning for specific species or landscapes can parallel nationalist movements. Like nationalist communities, whose only common denominator is the imagined ethnic or national identity, these global communities of mourners need only have one thing in common, namely that they imagine a psychological, private relation to what might be lost.

Where real responsibilities for loss are raised, they are immediately diffused. Environmental losses, just like terrorist acts, have become national security concerns – but so are, for Others, extractive energy policies and the use of drones. If we are mourning certain environments or certain people, how can we justify their destruction in other circumstances? The easiest way to deal with this situation is to replace complex kinship relations with simple or simplistic political boundaries between what is in our best interests and what is not. The private, psychological, and aesthetic emotions of grief are easily transferred to nationalistic emotions because, in the name of immediate reactions, the public is not allowed to take the time to grieve, to come to terms with loss and realize the complex kinship networks associated with losses. "The violence of ethnic conflicts ... does not tap into these relational processes," writes Gagnon (2004, 8) in the context of the Balkan wars. "Rather, its goal is to fundamentally alter or destroy these social realities." Such violence needs to project itself on its opponents, who are named terrorists or eco-terrorists.

In such a way, instead of grieving specific human or environmental entities such as clean water, a prairie landscape, or pronghorns, what Butler (2004, 39) calls the "loss of First World presumption," and bell hooks (1994, 32) names the "de-centering of Western civilizations" is taking precedence. Real, embedded social relationships are replaced by diffused pseudo-communities of vague, immediate, reflexive fear. Here, then, is the result of a political limitation of grievable entities and simultaneously a search for immediate action on the fears of other losses.

## Mourning and/as Healing

The real consequence of industrial alienation might be the qualitative shift from mourning for something that has been lost to mourning for something that might be lost in the future: anticipatory mourning. Just like some Inuit "were already imagining future losses, already experiencing levels of pain over what may come" (chapter 7, 172), what are we actually mourning when we feel grief over the loss of our environments? Increasingly, I would argue, we mourn the demise of our own species. In that way the global community does not mourn disappearing species in and for themselves, but rather as an omen of our own anticipated end. Anticipatory mourning, as Cunsolo points out, is something that many Inuit societies avoided, as anticipating future events might make them come true. However, there is another, more consequential qualitative shift involved. Mourning for our own end, of course, is not a social act anymore but a recoiling into oneself.

Theresa DeLeane O'Nell (1996) has shown how grief should lead to the realization that the loss of personal connections can be overcome or somewhat compensated by social connections to others because everybody incurs loss and needs pity. Similarly, Butler (2004, 30) argues that grieving "can be a departure for a new understanding if the narcissistic preoccupation of melancholia can be moved into a consideration of the vulnerability of others." The question of environmental relations is, in Butler's words (2004, 49): "What allows us to encounter one another?" It is, of course, a significant sign of a culture of non-places and alienation that we have to even ask that question. Todorov (1992) argued that "it is only by speaking to the other (not giving orders but engaging in a dialogue) that *I* can acknowledge *him* [or her] as subject" (132). We must, as bell hooks puts it, "genuinely *value* everyone's presence" (1994, 8). The aesthetification and commodification of relations that accompanies the development of deterritorialized lives denies such dialogues. A self-reflective awareness of living in distance and in difference – in exile – is useful to develop solidarity and traitorous identities (Plumwood 2000). However, "if a whole society consists of exiles, the dialogue of cultures ceases: it is replaced by eclecticism and comparatism, by the capacity to love everything a little, of flaccidly sympathizing with each option without ever embracing any" (Todorov 1992, 251). Dialogue, true encounters, and deep understanding of the Other, as Richard White showed masterfully in the *Middle Ground* (1991), is embedded in place and local action. It is (non-industrial) agrarian life, argues Orr (2003, 182), which "requires a patient and painstaking accommodation to the realities of life and death in the effort to husband the health and long-term productivity of particular places."

Such place-specific accommodations, of which mourning is one, may lead to healing. "Healing and place," postulates Gesler (2003, 1), "are inseparable." One purpose of mourning has to be that people's grief is healed, "to re-situate the deceased from the land of the living to their new habitation among the dead and to release the bereaved from the ashen netherworld of grief back to the realms of the living" (Baptist 2010, 297). Thus mourning is not necessarily intended to keep a memory of the dead forever. For example Taylor (1993, 655) argues that for lowland Amazonian cultures mourning accomplishes two tasks: "forgetting the dead as familiar persons, on the one hand, while still being able to think of them as social partners, on the other." Regardless of whether this is a generalizable specific function, the meaning of mourning lies in assigning living and dead their "right" place. Until this is done with certainty, mourners are not allowed to be social (St Pierre and Long Soldier 1995), or are prevented from enacting the social relationship in public, for example through name taboos (e.g. Wadley 1999; Thomson 1946). All participants are assigned their proper status of belonging (Geschiere 2005). When this is not possible, either because the dead are missing or because the proper rituals are not held in the proper landscape, no healing takes place (Taylor 1993; Baptist 2010). It becomes clear, then, that the healing of mourning is concerned with social relations, which have to be realigned. This can be hard work: "mourning is a 'work,' that is to say a *process*" (Taylor 1993, 674). The process is socially prescribed, depending on the socially accepted relationship to the dead. If the actual relationship was different, either closer than expected or more distant, that can lead to misunderstandings. Since mourning is a social practice – in contrast to grieving, which can be private – mourners usually have to conform to expectations or risk offending others (Fowlkes 1990; O'Nell 1996, 91–5; Huffstetter 1998, 129).

Just like refusing to grieve, refusing to end grieving denies one's past or continuing social responsibilities and leads to an ongoing trauma. Such trauma from loss – removal from landscapes, loss of culture, destruction of the ecosystem, etc. – can continue to haunt people for generations past their immediate impact, as long as "vivid recountings" keep it alive without a resolution (Schwarz 2008, 19–21). What is needed, as Yellow Horse Brave Heart and DeBruyn (1998) argue, is "the development of a spirituality that does not serve as a defense against experiencing painful affects. Rather, a healthy spirituality embraces the range of one's feelings – grief, shame, and pain to joy, pride, and resolve to maintain balance – in order to regain personal wellness and the power of community self-determination" (74). Judith Butler (2004) writes, "I do not think that successful grieving implies that one has forgotten another person or that something else has come along to take its place, as if full substitutability were something for which we might strive" (21). The goal of the process of mourning is to be able to readjust to loss, to adapt to the new social (and physical) landscape. Mourning for environmental loss in this sense is not geared to healing the landscape or the ecosystem, and it might not bring about actions to restore them. Mourning for environmental loss is pragmatic work, geared to adaptability and resilience more so than ethics of sustainability.

Sponsel (2001) thinks that an effective realization of spiritual ecology, the tenet that humans are an integral part of nature, "should make a positive difference in creating more adaptive human-environment interactions" (185). In light of the differences between environmental ethics and behaviour, and in light of the pragmatism of healing through mourning for environmental loss, "more adaptive" interactions, however, might simply mean that humans adapt better to changed environments. Unfortunately, I do not think that "a realization of the interconnections of all life will necessarily lead to a greater respect for all life" (Scharper 1998, 99). Similarly, mourning for the environment does not necessarily indicate a greater respect for life, and a greater respect does not necessarily lead to mourning. Mourning, as a social practice, simply reveals that environmental relations are social relations and, therefore, that if we want to create more positive relations with our environments we cannot do so in isolation from other social relations. To change our relations with our environments, then, we might need to change how we relate. The pseudocommunities of supermodernity stand in direct contrast to a meaningful relationship with, and a meaningful healing of our environments.

Ultimately, the incapacity to mourn is the indicator of not simply a loss of empathy for our environment, but also the loss of kinship relations – indeed of social meaning in general. That being the case, it might indeed be time to mourn for the loss of humanity.

## NOTES

1 I would like to thank the editors for a close reading and constructive criticism of this text. Their input allowed me to sharpen the argument and improve it considerably. All mistakes are mine.

2 On the link between military action and action against the environment, see also Vandana Shiva (2003), who writes about industrial agriculture as "a war against farmers" directly linked to terrorism.

## REFERENCES


Anderson, Benedict R. 1991. *Imagined Communities: Reflections on the Origin and Spread of Nationalism*. London: Verso.

Anderson, David G. 2000. *Identity and Ecology in Arctic Siberia: The Number One Reindeer Brigade*. Oxford: Oxford University Press.

– 2004. "Reindeer, Caribou and 'Fairy Stories' of State Power." In *Cultivating Arctic Landscapes: Knowing and Managing Animals in the Circumpolar North*, edited by David G. Anderson and Mark Nuttall, 1–16. New York: Berghahn Books.

Augé, Marc. 1995. *Non-Places: Introduction to an Anthropology of Supermodernity*. London: Verso.

Baptist, Karen Wilson. 2010. "Diaspora: Death without a Landscape." *Mortality* 15, no. 4: 294–307.

Barbeau, Marius. 1945. "Bear Mother." *Journal of American Folklore* 59, no. 231: 1–12.

Basso, Keith H. 1996. *Wisdom Sits in Places: Landscape and Language among the Western Apache*. Albuquerque: University of New Mexico Press.

Beatley, Timothy. 2004. *Native to Nowhere: Sustaining Home and Community in a Global Age*. Washington: Island Press.

Bennett, John W. 1996. *Human Ecology as Human Behavior: Essays in Environmental and Development Anthropology*. New Brunswick, NJ: Transaction Publishers.

Berkes, Fikret. 1999. *Sacred Ecology: Traditional Ecological Knowledge and Resource Management*. Philadelphia: Taylor and Francis.

Berry, Wendell. 1996. "Conserving Communities." In *Rooted in the Land: Essays on Community and Place*, edited by William Vitek and Wes Jackson, 76–84. New Haven: Yale University Press.

Braun, Sebastian Felix. 2008. *Buffalo Inc.: American Indians and Economic Development*. Norman: University of Oklahoma Press.

Buckland, Jerry. 2004. *Ploughing Up the Farm: Neoliberalism, Modern Technology and the State of the World's Farmers*. London: Zed Books.

Buffalo Field Campaign. 2006. "Buffalo Hunting Bill." http://www.buffalofieldcampaign.org/legislative/buffalohuntingbill.html.

Burton-Christie, Douglas. 2011. "The Gift of Tears: Loss, Mourning and the Work of Ecological Restoration." *Worldviews* 15: 29–46.

Butler, Judith. 2004. *Precarious Life: The Powers of Mourning and Violence*. London: Verso.

CampusActivism.org. 2004. "Yellowstone Buffalo – Slaughter Protest & Memorial Service." http://www.campusactivism.org/displayevent-447.htm.

Cooper, Gregory. 1996. "Aldo Leopold and the Values of the Native." In *Rooted in the Land: Essays on Community and Place*, edited by William Vitek and Wes Jackson, 150–60. New Haven: Yale University Press.

Cruikshank, Julie. 2004. "Uses and Abuses of 'Traditional Knowledge': Perspectives from the Yukon Territory." In *Cultivating Arctic Landscapes: Knowing and Managing Animals in the Circumpolar North*, edited by David G. Anderson and Mark Nuttall, 17–32. New York: Berghahn Books.

Danbom, David B. 2006. *Born in the Country: A History of Rural America, Second Edition*. Baltimore: Johns Hopkins University Press.

DeMallie, Raymond J. 1994. "Kinship and Biology in Sioux Culture." In *North American Indian Anthropology: Essays on Society and Culture*, edited by Raymond J. DeMallie and Alfonso Ortiz, 125–46. Norman: University of Oklahoma Press.

– 1998. "Kinship. The Foundation for Native American Society." In *Studying Native America: Problems and Prospects*, edited by Russell Thornton, 306–56. Madison: University of Wisconsin Press.

Eagleton, Terry. 1990. *The Ideology of the Aesthetic*. Oxford: Blackwell.

Ehrenfeld, David. 1996. "Pseudocommunities." In *Rooted in the Land: Essays on Community and Place*, edited by William Vitek and Wes Jackson, 20–4. New Haven: Yale University Press.

Feit, Harvey A. 2004. "The Enduring Pursuit: Land, Time, and Social Relationships in Anthropological Models of Hunter-Gatherers and in Subarctic Hunters' Images." In *Key Issues in Hunter-Gatherer Research*, edited by Ernest S. Burch Jr and Linda J. Ellanna, 421–39. Oxford: Berg.

Foster, John Bellamy. 2000. *Marx's Ecology: Materialism and Nature*. New York: Monthly Review Press.

Fowlkes, Martha R. 1990. "The Social Regulation of Grief." *Sociological Forum* 5, no. 4: 635–52.

Gagnon, V.P., Jr. 2004. *The Myth of Ethnic War: Serbia and Croatia in the 1990s*. Ithaca: Cornell University Press.

Geschiere, Peter. 2002. "Funerals and Belonging: Different Patterns in South Cameroon." *African Studies Review* 48, no. 2: 45–64.

Gesler, Wilbert M. 2003. *Healing Places*. Lanham: Rowman & Littlefield Publishers.

Gilkey, Langdon. 1993. *Nature, Reality, and the Sacred: The Nexus of Science and Religion*. Minneapolis: Fortress Press.
Grim, John A. 2001. "Indigenous Traditions and Deep Ecology." In *Deep Ecology and World Religions: New Essays on Sacred Ground*, edited by David Landis Barnhill and Roger S. Gottlieb, 35–57. Albany: SUNY Press.
Hallowell, A. Irving. 1926. "Bear Ceremonialism in the Northern Hemisphere." *American Anthropologist* N.S. 28, no. 1: 1–75.
Harrod, Howard L. 2000. *The Animals Came Dancing: Native American Sacred Ecology and Animal Kinship*. Tucson: University of Arizona Press.
hooks, bell. 1994. *Teaching to Transgress: Education as the Practice of Freedom*. New York: Routledge.
Huffstetter, Stephen. 1998. *Lakota Grieving: A Pastoral Response*. Chamberlain: Tipi Press.
Ingold, Tim. 2000. *The Perception of the Environment: Essays in Livelihood, Dwelling, and Skill*. New York: Routledge.
Jackson, John Brinckerhoff. 1994. *A Sense of Place, a Sense of Time*. New Haven: Yale University Press.
Jackson, Michael. 2013. *Lifeworlds: Essays in Existential Anthropology*. Chicago: University of Chicago Press.
Jaffee, Daniel. 2007. *Brewing Justice: Fair Trade Coffee, Sustainability, and Survival*. Berkeley: University of California Press.
Jameson, Fredric. 1998. *The Cultural Turn: Selected Writings on the Postmodern, 1983–1998*. London: Verso.
Katz, Eric. 2000. "Against the Inevitability of Anthropocentrism." In *Beneath the Surface: Critical Essays in the Philosophy of Deep Ecology*, edited by Eric Katz, Andrew Light, and David Rothenberg, 17–42. Cambridge, MA: MIT Press.
Krech, Shepard, III. 1999. *The Ecological Indian: Myth and History*. New York: W.W. Norton.
Leopold, Aldo. 1991. "The Ecological Conscience [1947]." In *The River of the Mother of God and Other Essays by Aldo Leopold*, edited by Susan L. Flader and J. Baird Callicot, 338–46. Madison: University of Wisconsin Press.
Lévi-Strauss, Claude. 1969. *The Elementary Structures of Kinship*. Second Edition. Boston: Beacon Press.
McKinley, Robert. 2001. "The Philosophy of Kinship: A Reply to Schneider's Critique of the Study of Kinship." In *The Cultural Analysis of Kinship: The Legacy of David M. Schneider*, edited by Richard Feinberg and Martin Oppenheimer, 131–67. Urbana: University of Illinois Press.
Mills, Antonia, and Richard Slobodin, eds. 1994. *Amerindian Rebirth: Reincarnation Beliefs among North American Indians and Inuit*. Toronto: University of Toronto Press.
Nesper, Larry. 2002. *The Walleye War: The Struggle for Spearfishing and Treaty Rights*. Lincoln: University of Nebraska Press.

Niezen, Ronald. 2009. *The Rediscovered Self: Indigenous Identity and Cultural Justice*. Montreal: McGill-Queen's University Press.

O'Nell, Theresa DeLeane. 1996. *Disciplined Hearts: History, Identity, and Depression in an American Indian Community*. Berkeley: University of California Press.

Orr, David W. 2003. "The Uses of Prophecy." In *The Essential Agrarian Reader: The Future of Culture, Community, and the Land*, edited by Norman Wirzba, 171–87. Washington: Shoemaker and Hoard.

Plumwood, Val. 2000. "Deep Ecology, Deep Pockets, and Deep Problems: A Feminist Ecosocialist Analysis." In *Beneath the Surface: Critical Essays in the Philosophy of Deep Ecology*, edited by Eric Katz, Andrew Light, and David Rothenberg, 59–84. Cambridge, MA: MIT Press.

Popper, Deborah Epstein, and Frank J. Popper. 1987. "The Great Plains: From Dust to Dust." *Planning* 53, no. 12: 12–18.

Rana, Matthew David, and Anthony Marcellini. 2012. "Notes toward a Non-anthropocentric Social Practice." http://www.artpractical.com/feature/notes_toward_a_non_anthropocentric_social_practice/.

Rappaport, Roy A. 1999. *Ritual and Religion in the Making of Humanity*. Cambridge: Cambridge University Press.

Riley, Robert B. 1997. "The Visible, the Visual, and the Vicarious: Questions about Vision, Landscape, and Experience." In *Understanding Ordinary Landscapes*, edited by Paul Groth and Todd W. Bressi, 200–9. New Haven: Yale University Press.

Rose, Deborah, Diana James, and Christine Watson. 2003. *Indigenous Kinship with the Natural World in New South Wales*. Hurtsville: NSW National Park and Wildlife Service.

Sahlins, Marshall. 1976. *Culture and Practical Reason*. Chicago: University of Chicago Press.

Schneider, David M. 1980. *American Kinship: A Cultural Account, Second Edition*. Chicago: University of Chicago Press.

– 1984. *A Critique of the Study of Kinship*. Ann Arbor: University of Michigan Press.

Schwarz, Maureen Trudelle. 2008. *"I Choose Life": Contemporary Medical and Religious Practices in the Navajo World*. Norman: University of Oklahoma Press.

Sharp, Henry S. 2001. *Loon: Memory, Meaning, and Reality in A Northern Dene Community*. Lincoln: University of Nebraska Press.

Shiva, Vandana. 2003. "Globalization and the War against Farmers and the Land." In *The Essential Agrarian Reader: The Future of Culture, Community, and the Land*, edited by Norman Wirzba, 121–39. Washington: Shoemaker and Hoard.

Sponsel, Leslie E. 2001. "Do Anthropologists Need Religion, and Vice Versa? Adventures and Dangers of Spiritual Ecology." In *New Directions in Anthropology and Environment: Intersections*, edited by Carole L. Crumley with A. Elizabeth van Deventer and Joseph J. Fletcher, 177–200. Lanham: Altamira Press.

St Pierre, Mark, and Tilda Long Soldier. 1995. *Walking in the Sacred Manner: Healers, Dreamers, and Pipe Carriers – Medicine Women of the Plains Indians*. New York: Simon & Schuster.

Thomson, Donald F. 1946. "Names and Naming in the Wik Moŋkan Tribe." *Journal of the Royal Anthropological Institute of Great Britain and Ireland* 76, no. 2: 157–68.

Thornton, Thomas F. 2008. *Being and Place among the Tlingit*. Seattle: University of Washington Press.

Todorov, Tzvetan. 1984. *The Conquest of America*. New York: HarperCollins.

Van der Kroef, Justin. 1991. "Some Head-Hunting Traditions of Southern New Guinea." *American Anthropologist* N.S. 54: 221–35.

Wadley, Reed L. 1999. "Disrespecting the Dead and the Living: Iban Ancestor Worship and the Violation of Mourning Taboos." *Journal of the Royal Anthropological Institute* 5, no. 4: 595–610.

Warren, Karen J. 2000. *Ecofeminist Philosophy: A Western Perspective on What It Is and Why It Matters*. Lanham: Rowman and Littlefield Publishers.

White, Richard. 1991. *The Middle Ground: Indians, Empires, and Republics in the Great Lakes Region, 1650–1815*. Cambridge: Cambridge University Press.

Wishart, Robert P. 2004. "A Story about a Muskox: Some Implications of Tetlit Gwich'in Human-Animal Relationships." In *Cultivating Arctic Landscapes: Knowing and Managing Animals in the Circumpolar North*, edited by David G. Anderson and Mark Nuttall, 79–92. New York: Berghahn Books.

Yellow Horse Brave Heart, Maria, and Lemyra M. DeBruyn. 1998. "The American Holocaust: Healing Historical Unresolved Grief." *The Journal of the National Center on American Indian and Alaska Native Mental Health Research* 8, no. 2: 60–82.

# 4 In the Absence of Sparrows

HELEN WHALE AND FRANKLIN GINN

## Introduction

Colourful flocks and feathered shapes make their way across a page. They are moving yet motionless, left to hang forever from a branch, their trajectory of flight suspended in time. The illustrations are from Hume and Walters's 2012 *Extinct Birds*, said to be the first study to document every avian species' demise to have occurred over the last thousand years. As this volume demonstrates, extinction haunts our time as never before. The extinction that pervades popular nature-writing and the environmental sections of our newspapers, and which confronts us through screens, is not about the past. The blurry shot of a species on the brink (*Guardian* 2012), the secret blooming of a rare flower (Challenger 2011), the "last chance to see" (Adams and Carwardine 1991): these things instead suggest that the nagging *presence* of that which threatens to disappear – a disappearance both happening and perpetually threatening – captures the imagination even more.

In times of both abundance and scarcity, birds have been a source of inspiration and fascination for the human imagination. In the UK, home of the Royal Society for the Protection of Birds (RSPB), now the largest conservation charity in Europe, bird-watching and feeding birds are ever popular. Around 50 per cent of the British public are now said to provide food for the birds visiting their yards (Cammack et al. 2011). At the same time, many bird species are in severe decline. At the most recent count, fifty-two of 246 assessed species were coded "Red" on the

UK list of birds at greatest risk of national extinction (RSPB 2009). This Red List has itself attracted popular interest. In 2007, the journalist Charlie Elder travelled the UK with a checklist of Red birds, determined to see and document those species under threat. More than just appealing to the "twitching" sensibility (Lorimer 2007), however, Elder's (2009) *While Flocks Last* poses some important questions: while they may be rapidly disappearing from the UK, he observes, many of the Red List species are not at risk of global extinction. What does the presence or absence of a particular bird *in a particular place* mean, Elder appears to be asking – and why should we care? Later, as a mere aside, he offers a thought: Birds and their songs, he suggests, "evoke a vivid sense of place" (2009, 53). If birds are evocative of place, what happens when birds disappear? Is it merely a question of loss, or is there more happening? How do we experience that which is absent? This chapter addresses such questions, attending to the decline and conservation of one species in particular: the house sparrow (*Passer domesticus*) in London, UK.

The decline of the house sparrow in the UK is a relatively recent phenomenon. From around 13 million pairs in the early 1970s, numbers are now estimated to stand somewhere between 2.1 and 3.7 million: a 71 per cent drop in the national breeding population from 1977 to 2008 (Shaw et al. 2008; RSPB 2012). While both urban and rural populations have changed, it is the decline in city centres that has attracted the most attention. Central London, Edinburgh, and Glasgow appear to be reporting near-extinction locally (RSPB 2012). In 2002, in the midst of this documentation of its disappearance, therefore, the house sparrow made its way onto the UK Red List.

Potential causes of sparrow decline have been the subject of extensive research (Summers-Smith 2003). As their name suggests, house sparrows have long been associated with human settlements and ways of life (Shaw et al. 2008). Theories put forth for the urban decline have therefore largely been anthropogenic: loss of nesting sites, lack of invertebrates for chicks owing to increased pollution, changes in particulate pollution following the switch to unleaded petrol, and interference to chick-rearing from traffic noise (although increased predation by other birds is another suspect) (Summers-Smith 2003; Shaw et al. 2008; Tuffrey 2012). The exact ecological underpinnings remain complex and contested; however, many sources suggest that a combination of the above factors is likely (Summers-Smith 2003). Research and conservation of house sparrows is ongoing.

In 2009, in conjunction with borough councils and the Royal Parks, the RSPB launched the London House Sparrows Parks Project (LHSPP). Grassland plots or "meadows" of three different types (long grass, native wildflower, and wildlife seed) were cultivated in twenty-five public parks across London. The RSPB hoped that, by increasing potential invertebrate numbers and providing seed, the plots would encourage house sparrows and other birds. In addition to the staff and researchers in charge of monitoring invertebrate and pollinator numbers, the LHSPP had volunteers, each of whom was assigned a local plot to survey, recording all birds landing in the designated area during a forty-five minute timeslot each month. To provide a control, adjacent plots of unseeded short grass were also surveyed.

This chapter does not concern itself directly with the ecological complexities of house sparrow decline. Through interviews with those involved with the LHSPP, it explores how house sparrow decline can be understood in terms of mourning, landscape, and experiences of absence and presence. We argue, in essence, that the absence of sparrows is not simply about a loss. Rather than being simply a subtraction, the absence of sparrows creates something new: haunted and spectral sparrow places.

## Place, Conservation, and Disappearing Birds

Seeking to broaden understandings beyond notions of a space filled with objects, geographers have historically underlined the importance of how ideology and meaning is encoded into the landscape (Cosgrove and Daniels 1988). More recently, within geography and anthropology much work has drawn upon Maurice Merleau-Ponty's writings (1962 [1945]) on phenomenology and perception to rethink place as dwelling and being-in-the-world (Feld and Basso 1996; Ingold 2000; Rose and Wylie 2006). Rather than pre-existing the individual, place here is constituted by and constitutive of embodied engagement (Rosenberg 2007). That is to say, place arises through presence – both of the body and of the multi-sensory landscape that surrounds it. In phenomenological understandings, therefore, the body acts as a sensing vehicle "fusing" self, time, and world, and thus enabling space and landscape to become place (Wylie 2009). For Feld (2005), place is not only landscape seen, smelled, heard, touched, and tasted but also landscape *remembered* through these very senses. As he argues, "senses make place" (179). Place might not only arise through the body being *in* place, in full sensory, sensing mode, then, but also through the evoking of "a feeling that we ... know what it is like to 'be there'" (Cresswell 2004, 7–8). As particular sense experiences come to be associated with particular times and places, sensory feeling entwines itself with emotional feeling; memory and sense congeal in place, and place congeals memory and sense (Casey 2000). This notion of place owes much to what might be called the rootedness of sense and memory. As Tuan (2004) sees it, however, the very notion of rootedness that place can connote is more complex than it first appears. Sense of place in his view emerges as a projection of the human psyche, an unconscious attempt to stabilize place against a backdrop of unavoidable change – change in terms of both self and time. Accordingly, then, the trouble for Tuan is that "place must stop changing for a human being to be able to grasp it and so have a sense of it" (2004, 45).

This is of course an impossible demand. Non-representational and mobilities geography have stressed the fleeting nature of place, and its complex relation with memory (Laurier and Lorimer 2012). One important feature of such work has been the attempt to portray places as having complex, non-linear temporalities (Dodgshon 2008). Places are not just congealed by memory, they do not just come to weigh down experiences of place; rather, new associations are simultaneously made through every experience, resulting in "an articulation of presence as 'the tangled exchange of noisy silences and seething absences'" (Gordon in Thrift 1999, 316–7). Place, in this sense, is a slippery *becoming* that "can never be completed" (Thrift 1999, 317). For Thrift, to get a grip on an "ecology of place" is to recognize that "places are 'passings' that 'haunt' us," forever slipping out of reach (ibid., 310). To speak of a sense of place, then, might be to speak of the conjuring up of such hauntings.

Indeed, recent spectral geographies have sought to conceptualize place as haunted and/or haunting (Maddern and Adey 2008). Much writing has concerned encounters with the city, where material traces of past lives and industries within the landscape speak of the absent, the ghostly, and the disappearing (Edensor 2005; Ginn 2013a), but also other spaces of everyday ruin (DeSilvey 2007; MacDonald 2013; and a 2013 special issue of the journal *Cultural Geographies* 20: 423–539).

John Wylie has explored absence through an account of his own experiences walking along a Cornish clifftop. Coming across a set of memorial benches, he finds himself witness to "blind-spots," as his own ways of seeing the landscape meet – yet fail to coincide with – the perspectives of absent loved ones. It is in this sense, he argues, that notions of absence come to problematize the supposed "co-presence of self and landscape" described in phenomenological accounts (Wylie 2009, 278). Yet, as Wylie sees it, spectral geographies themselves might also have neglected ways in which *absence itself* can enter into – and disrupt – the frame of perception (ibid., 279). In his view, the real challenge to a phenomenological understanding of landscape is not just unearthing memory, "bringing-to-presence [and] making the invisible visible," but rather coming to see "absence at the heart of the point of view" (ibid.).

Wylie's writing on absence draws heavily upon the work of Jacques Derrida, for whom "the phenomenal form of the world itself is spectral" (Derrida 1994, 135). For Derrida, place is something conjured and so can only be conceived within terms of "hauntology," an understanding of history and temporality in which the present is haunted both by the past and the future. Rather than an ontological matter of what *is*, the present is caught between what *was* and what *might be*, throwing into question the "contemporaneity of the present to itself" (ibid., xix). Central to Derrida's deconstruction of notions of presence and absence is the supplement (Royle 2003), which marks – adds to – an absence, whilst simultaneously attempting to make up for this absence. In this sense in an apparently illogical twist of logic an absence is simultaneously both revealed *and* filled. The supplement might appear to be both presence and absence but as Derrida tells us, it is actually neither. In Wylie's Derridean reading, pasts, presents, and futures – simultaneously timespaces *and* places – come to interrupt and erupt into one another in a supplemental relation (Wylie 2007). Accordingly, Wylie suggests that "haunting is a pre-requisite to place"; it appears "as a sudden and displacing *punctum* of pasts and presents" (ibid., 181–2).

Where Wylie suggests that spectral geographies have largely been concerned with a "bringing to light [of] things previously hidden or lost" (Wylie 2009, 279), a similar process of making the invisible visible has been identified within the theory and practice of wildlife conservation. Geographers interested in nature, or with entanglements of human and non-human lives, have begun to turn their critical eye to the representational practices focused on the rare and the declining. Studies have examined the modes of framing such as monitoring, surveying and recording, and how they might seek to bring presence to light. Lorimer (2008, 380), for example, has identified such a pattern within RSPB corncrake conservation, arguing that particular modes of representing "sought to disentangle the mystery of the corncrake." In his urban political ecology of black redstart and water vole conservation, Hinchliffe (2008, 88) argues that "nature conservation is concerned with revealing presence and rendering the present eternal." This very process of present-ing, as he sees it, points to a problematic at the heart of nature conservation. That is, as he sees it, while nature "has to be present" to be saved, many species resist the reductive binaries of absence and presence, defying the static spatialities of modes of recording, protecting, and sheltering (ibid.).

In general, endangered species are brought into view not only through processes of recording and monitoring, but more widely through archives and public campaign materials: photographs, critter cams, leaflets, and Red Lists. Yusoff's (2011) examination of what she calls the "aesthetics of loss" reveals a subtlety in this very process of "making present." In bringing the "absent one" into the field of vision, she argues, visual materials only render species *more* spectral, for the creatures "are not present in this encounter as themselves ... but as a kind of haunting configured around a profoundly human sensibility" (Yusoff 2011, 8). In seeking to "connect" us, such prosthetics merely underline the absent other's distance and disconnect (Candea 2010; Ginn 2013b). For Yusoff, presence and absence are not only dialectically engaged, but "part of a relation with possibility – the ability of the subject to be sensible to the constitution of the other" (2011, 3). In this sense, she contends, conservation practice seeks to replace the melancholy of absence with an ever-sought-after encounter of presence.

And so, the melancholy of absence has begun to receive critical attention. Ryan's (2009 and chapter 5) works on threatened Australian flora are some of the few to have addressed it more directly. For him, the "disappearance of species, as a complete extinction or as a gradual decline in occurrence, has emotional and aesthetic consequences that entail dealing with loss, silence and absence" (Ryan 2009, 51). He considers the double presence of species in the world – both as listed taxonomies, the "reified aspect of the species," and as lively material things "on the ground and in actuality" (Ryan 2009, 53). For Ryan, then, absence in the context of species loss emerges as twofold. Yet both Ryan and Yusoff contend that a species, as a taxonomic representation, cannot be grieved. Rather, sadness hinges upon a relational conception of human connections with the more-than-human world(s); "one mourns the loss of self, the loss of the other and the loss of the connectivity" (ibid., 75).

Mourning in this Derridean sense is not the traditional psychoanalytic return to "normal" after a loss, but a diminishment of the prospects for becoming. Derrida suggested that to internalize someone in memory denies their independence, it erases their autonomy, but on the other hand not to remember them means that we lose them completely. In the traditional psychoanalytic understanding of mourning, Derrida argues, we end up losing the deceased a second time, as they are acknowledged, but then put away again "in us" – their otherness is removed. We have to mourn, but if we fully internalize the other we are no longer faithful to that which we mourn. The solution, Derrida suggests, is not to transcend or reconcile this tension, but to recognize it, and to retain a sense of the irretrievability of the lives of others. Retaining a sense of the irretrievability of the past "challenges the survivor to be faithful not just to ... memory but also to singularity and alterity" (Kennedy 2007, 118). Allowing absent creatures to circulate as ghost, spirit, memory, or material trace sits paradoxically between an attempt to keep hold of that which has passed, and the recognition that it is impossible to do so faithfully without in some way diminishing that which is mourned.

In contrast to the positive possibilities of conservation practice, the idea of the absent or the lost has long been a poignant subject for nature and landscape writing. Perhaps as a result of their (seemingly) obvious presence yet simultaneous flightiness, birds have attracted particular attention. In the revered *Sand County Almanac*, Leopold (1949) articulates the grief felt in the aftermath of the extinction of the passenger pigeon, a species once present in billion-strong flocks. With the passing of those who lived to experience the pigeons, he suggests, the memory of the birds will come to lodge itself instead in the trees and landscapes in which they once abounded. Another piece in the same book, "Marshland Elegy," takes this further. Here grief is visible, imbued in the landscape itself: the "sadness discernible in some marshes arises, perhaps, from their once having harboured cranes" (Leopold 1949, 97). Without their feathered former inhabitants, we read, the marshes are left "adrift in history" (ibid.). Leopold's title itself is revealing: it is the place – the marsh, and not the crane – for which we grieve. In the absence of the cranes themselves, absence and grief arise through perception of landscape, but also from the *knowledge* that it was once a place of cranes.[1]

Birds have not only been associated with particular places, but also with particular times in human history. For example Price contends that in the context of rapid urbanization and change, the loss of the passenger pigeon flock came to stand for a wider loss: that of the rural landscape and way of life. Her archival research uncovers elegies shot through with nostalgia from those who encountered (and often hunted) pigeons while growing up in the country: "When young Mershon had gone out pigeon hunting ... he'd gone off to shoot a bird that soon would be as irretrievable as his childhood" (Price 1999, 47). In contrast to Leopold, however, whose elegiac writing mourns the loss of the pigeons as pigeons, Price's research suggests that an expressed nostalgia for pigeons spoke more of human responses to change than of the birds themselves.

Of course, Leopold himself never experienced flocks of passenger pigeons darkening the sky or weighing down trees: they were already on their way out when he was born. As Heise explains, then, there is also a tendency for extinction literature surrounding birds to "foreground feelings of loss and mourning, even on the part of individuals who have never themselves laid eyes on the bird" (2010, 66). In addition to Elder's (2009) *While Flocks Last*, recent nature writing has seen authors documenting their own journeys in search of rare species, as they attempt to witness for themselves that which threatens to disappear and "tumble beyond reach" (Challenger 2011, 58). Seeking out the elusive corncrake on a Scottish isle, for example, the poet Kathleen Jamie feels "denied one of the sounds of summer that all our forebears would have known" (2005, 98); for her, the corncrake-less mainland becomes a "diminished place." Indeed, Lorimer's (2008, 392) work has also pointed to this potential for the corncrake to emerge as "an aurally aesthetic, evocative and poetic fragment of the landscape that triggers the popular imagination ... a symbol of a shattered pastoral past."

Such cultural-historical undertones are raised more explicitly in Heise's (2010) own delineation of literature responding to the decline and (likely) extinction of the ivory-billed woodpecker. She suggests that the bird became entwined with a yearning for a way of life specific to a particular place, even and especially on the part of those who – like Leopold with the pigeons – had never actually experienced it. In the narratives of extinction, Heise argues, "part of human identity and culture itself seems to be lost along with the disappearance of a non-human species" (2010, 69). It seems, paradoxically, that as decline rendered them materially absent, the passenger pigeon, the ivory-billed woodpecker, and the corncrake became *more* "present" in the collective consciousness, as aesthetic objects of mourning around which nostalgic longings for a particular cultural-historical landscape and place-based identity could crystallize (Heise 2010). That is, in their very disappearance these particular birds seem not to disappear. Rather, they linger on: in the imagined landscape, but also through vocalized human desires for what such landscapes, in their very disappearance, have come to represent. Conservation practice, in rendering species either present or absent, is predicated on future nostalgia for lost species and their companion worlds.

In what follows, we address how place might be understood as a disruptive unsettling, both arising through and resulting in complications of what we understand as presence. We suggest that the absence of sparrows is not just about a loss; rather, the ways that sparrows circulate – as mnemonics, on signs, as half-heard trills, half-glimpsed flutters in memory, imagination, or place and, crucially, in their absent presence – are generative of place. The chapter draws on fieldwork conducted by one of the authors, Whale, in London, UK in 2012. Twelve bird enthusiasts were recruited through the RSPB and an online networking site for London birders: seven were volunteers on the House Sparrows Parks Project, four were RSPB staff members, and one was a freelance ornithologist. All but one of these interviews took place outside. The practice of walking is well suited to exploring notions of place, self, memory, and haunting, both alone and with participants (Wylie 2009; Butler 2006; Kusenbach 2003), as well as nature experiences (Waitt et al. 2008). Accordingly, most interviews took place while walking around the site of sparrow plots. Our analysis suggests that in many ways the House Sparrows Parks Project was as much about mourning as about conservation, and that something of the paradoxical nature of mourning haunts conservation practice itself. The project was both faithful to that which was absent (the house sparrow), but also refused to admit that the absent sparrow really was totally absent. Conservation practice encompasses the preserve, of course, but also, we suggest, hinges on a less-acknowledged recognition of the impossibility to preserve fully. Conservation, then, is always already mourning its own failure.

Although this story is faithful to the testimony of interviewees it is also our own story and perhaps, therefore, one that project volunteers might not recognize. We hope it is clear, nonetheless, that we empathize with their practice and their labour, and that we share with them a love of sparrows in all their guises. The rest of this chapter evokes the absence of sparrows in five ways: meadow, hedge, flock, childhood, and London.

## Meadow

By the summer of 2012 the London House Sparrows Parks Project was over and the sparrow meadows themselves were starting to disappear. Several interviewees were returning for the first time since it ended and, wandering around their sites, they remarked on changes. They spoke of daisies, poppies, and sunflowers, long grass (often still present), thistles. Some spoke of fences and signs, now long removed or stolen, separating the trial areas from the surrounding park. And yet as Sarah, the House Sparrows Parks Project co-ordinator, pointed out some of these signs remained. For the volunteers, the sign's role is simple: to provide information and raise public awareness. In this first encounter with the sparrow, then, it is this seemingly simple role of the sign that we wish both to explore and challenge.

What happens when one looks at this sign? To us, what seems clear is that it is not only the person looking who is looking: the sparrow meets one's gaze, commanding that you draw your attention in. This very drawing in of the gaze – which is met by the sparrow – encourages one to look more closely at the details of the sign,[2] which in turn precipitates

Figure 4.1 | Sign marking the "first encounter" with a sparrow.

another response: the casting out of the gaze to take in the landscape that both surrounds and lies beyond the sign. As Yusoff (2011, 9) has identified, it is this "bringing into view" of the face of the absent one that prepares the human sensibility for the *possible*; for meeting with "the anticipated one." As Butler (in ibid., 5) contends, "the senses are part of any recruitment effort." Certainly as this larger-than-life, more-than-visible sparrow watches, we are inclined to scan the meadow for its "material" counterpart, looking and listening. Yet might it be possible to see something more here – or rather, something less? That is, through its words as well as its imagery, the sign is also telling us of sparrow decline; we simultaneously look for *absence*. Something strange is happening, then: one is both looking for (searching) and looking at (projecting). The sparrow on the sign is not the final resting stop for the gaze; rather the sign acts as a medium to direct the senses to both witness sparrows *and* their absence. The sign does something more than gather in the gaze: it seems to radiate outwards. To elaborate, we will return to Sarah. When we met at the RSPB office, the obvious question was: what was the LHSPP trying to do? She explained it by saying that, "with the communications part of the project, I think it might just sort of make people notice, and realise there's something missing." What Sarah seems to be suggesting here is that rather than just making absent sparrows present, the LHSPP strove to direct the public gaze towards absence, to make absence itself something visible. It should be made clear that Sarah is not talking exclusively about the meadows and the signs; communications also included public events, leafletting, and widespread publicity. With Sarah's words in mind, though, what we can suggest is that the sign – in offering a face *and* a story of decline – goes beyond "making present" to enact something more akin to Leopold's (1949) description of the crane(less) marsh. As a space supposedly occupied by sparrows, as the sign tells us, the meadow inherits a history full of sparrows, of sparrow-fullness, which immediately sets it at odds with its own (apparent) sparrow-less present.

Furthermore, in this present, the meadow appears to be lost in time. The meadow lingers between a conjured sparrow past and an imagined sparrow future; indeed the RSPB describes the past as something to work towards in the future. Recall from our earlier discussion that this tension between fidelity to the past and the impossibility of continuing on, ever-faithful to that which has vanished, is a key part of the work of mourning. The meadow, through this work of mourning, becomes a haunted – and haunting – sparrow *place*. Experiencing the meadow alongside the sign, we look for and at both presence and absence, and yet, it seems, neither is really there.

## Hedge

For the volunteers, of course, the meadows were also places of looking. Unsurprisingly, conversations quickly turned to birds. Interviewees were keen to describe what they had seen. Gulls, pigeons, crows, the odd exciting rarity; no one recorded a house sparrow, save Barbara, at the very beginning of the project. They were looking, but in the absence of sparrows.

And yet only one or two of the volunteers expressed sadness about a lack of sparrows. Instead many offered likely reasons for this absence: dog walkers, soccer games, disturbance from works or roads. Furthermore, as Michael was keen to point out, "sparrows don't come out into the open in the way that they used to." The plots themselves, at least when the volunteers were watching them, were not necessarily conducive to sparrow presence. Chris said that, "behind the plot, there's hedgerows, and when I was sitting there not seeing anything I was ... I'd think, if I was a bird I wouldn't come into this ... plot of longish grass; I'd live in the hedgerow behind." Chris often didn't record birds of any kind on his plot.

While the surveying process requires the bird to make itself known, Chris tries to picture where birds might be hiding. He balances his not seeing anything by taking an imaginative leap into small bird sensibility. In other words, Chris wants to say that a lack of visible birds in the plot does not necessarily denote absence from the park; as Hinchliffe (2008, 91) contends, absence "from records does not necessarily mean 'not present.'" If the sign, as examined in the preceding section, transforms the meadow into a clearly demarcated place where sparrows are absent, Chris's bird-watching sensibility, on the other hand, works to undo this. He engages not only with the plot itself, but also with the surrounding landscape and the possible, hoped-for presence of sparrows in the hedge ("if I were a bird I'd live in the hedge behind," he says). Small birds, it seems, are no more obviously "present" in the hedgerow than they are in the haunted meadow. Further, the hedge does not easily present itself to sensory exploration; Chris does not look, and instead he imagines. What Chris's comment is pointing towards, we suggest, is that being on the lookout for sparrows – and experiencing an (apparent)

Figure 4.2 | One of the "hidden" spaces of sparrows.

absence of sparrows – involves not only watching, but a feeling of being watched: maybe.

To see how this works, we turn to Jim, a freelance ornithologist and bird surveyor, who spoke about the difficulties of conducting sparrow counts. "You may not be aware of them," Jim said, "unless you see them fly, or you see them moving around, but in a sort of dense privet hedge, even if you walk up to them, like a lot of birds, they'll shut up, because of safety, they know you're there." Sparrows today, it seems, like lurking in hedges. Given this, Jim is explaining, not only are they difficult to survey visually, but auditory counts can be unreliable as well. Even for the trained birder, sparrows resist any easy numerical representation. Looking more closely at Jim's words, however, we uncover something stranger than the practical difficulties of recording non-human presence. Jim is saying that as surveyor he is not always aware of sparrows hiding in the hedge. Yet at the same time, he is *aware that he might not be aware*. For the surveyor, gifted with this double, framed awareness, the sparrow is neither simply absent nor simply present. Rather it is something possible; something that might be there and might be watching but which might not make itself known. The sparrow itself, meanwhile, becomes the knowing subject.

How might we understand this sparrow, then, and the possible sparrows that interviewees referenced when attempting to explain their sparrow-less plots? Here we might turn to Wylie's (2007, 172) reading of Derrida. The spectre, Wylie explains, "is invisible, but it watches, it is the unseen seeing." As Derrida (1994, 136) himself wrote: "one feels oneself looked at by what one cannot see." It has been our contention in this chapter that the act of looking for sparrows is both contingent upon and unsettled by a sparrow that always might be lurking somewhere in the foliage. Based on the experiences of interviewees, we might thus come to see absence in the context of sparrow decline as a hiding, rather than a loss. Accordingly, we might also come to feel that places "are always already haunted insofar as 'they' [spectres] are everywhere where there is watching" (Wylie 2007, 172). Loss then should not be seen as the subtraction of sparrows from the world, but as a displacement, as a shift into less certain terrain and a production of haunted place.

## Flock

In the last section, we quoted Michael, who said that sparrows "don't come out into the open in the way that they used to." Others spoke of the "poignancy" (Barbara, Maria) of seeing a single individual – a sparrow without other sparrows, a sparrow out of the flock. While the sparrow on the RSPB sign (figure 4.1) we saw earlier is a single individual – typical of conservation campaign literature – what some interviewees seemed to be implying was that sparrows *should come in flocks*. Like the passenger pigeon, Jim and Julia were both keen to point out, they are a colony species; breeding in a flock is desirable for survival. Not only therefore does the apparent absence of a sparrow seem to speak of other times, other places, but also about a loss of something else: of specific ways of being-sparrow, of absent flocking, nesting, and movement behaviours.

The sight of a single sparrow also seems to speak of – or perhaps *for* – other sparrows; in the context of decline, it supplements the house sparrow as species, opening up a space for an absence to be *made present*. There is something particular to the house sparrow as a species – or at least to our human conceptions of what it *is* to be a sparrow – which means that witnessing one involves something more than merely ticking a name off a list. Sparrows, like some of the species Hinchliffe (2008) identifies, resist a straightforward binary of absence and presence. Yet this is not only when presence might be fleeting or numbers difficult to determine, but even when they seem to present themselves directly to the perceiving eye.

With this "single sparrow" in mind let us turn to Barbara. Here she is describing a rare encounter with a flock of sparrows at a garden cafe in Essex. Helen asked her, "When you do visit a place where there are lots, what's it like, to experience that?" Barbara responded: "Well, it's ... it's quite nostalgic really. I remember being at Hyde Hall which is one of the RHS show gardens, and there were lots of sparrows flocking around the cafe there where we were, sneaking the crumbs and so on, and it's just, yes it's quite a nostalgic sort of sight, you sort of remember what it was like having these little brown things everywhere. And yeah, you sort of miss them, really." Seeing a flock of sparrows at a cafe table creates a feeling of nostalgia in Barbara. Unlike those who mourned the loss of the passenger pigeon (Price, 1999), however, it is the supposedly direct and immediate sight of the birds that Barbara describes. That is, she seems to yearn for something that appears still to be there; their present "presence" testifies to a wider loss or absence of the very spectacle that appears before her. That is, being witness to this sight allows memory to do its work, moving her between her past, in which the "little brown things" were everywhere, and her present, in which sparrows are more generally absent. Earlier, we saw how invisible "possible sparrows" might be understood in the light of the spectral. Seeking to understand Barbara's experience, as well as the experience of witnessing a single sparrow, we suggest, we might again turn to Derrida, and come to see that "the spectre is visible, it appears, but 'there is something disappeared, departed in the apparition itself as reapparition of the departed'" (Wylie 2007, 172) When it comes to sparrows today, it seems, neither presence nor absence alone can fully describe how they make themselves known in the world. What is missing, then, is not just birds. What is missing is sparrows that are like what sparrows used to be like (they used to flock;

they no longer flock). The realization is that what is absent, and what may be hiding, is not the same as itself. We can see how, then, the very object of mourning is unstable, shifting, and continuing to change.

## Childhood

Many LHSPP volunteers had been regular visitors to a particular spot in London and had been keeping records of declining numbers. For most of them, however, an absence of sparrows had only been noticed when visiting elsewhere – often when returning to the family home in another city or the countryside. When discussing their recent encounters with sparrows in London, a certain sense of comfort at something "still being there" emerged: chirruping was described as "reassuring" (Daniel); "comforting" (Sarah); "homely" (Michael). As well as confirming the continued lively presence of the house sparrow as species, it seemed that many of participants delighted in sparrow presence in that it transported a sense of home to the urban environment: it made the landscape a home away from home.

Later, as attention turned away from direct surroundings, many interviewees began to reminisce: feeding sparrows at home with the family, counting them as a child, sometimes even firing at them with catapults. More often than not, though, it was sparrow *sounds* that were described as triggering memories of childhood. We begin with Charlie, who wants to explain why sparrow sounds are important to him. "It's not necessarily the prettiest of calls or songs, if you can call it a song, but there is something ... and perhaps it is just going back to childhood. I was so used to hearing that, that was just the sound of ... well I suppose the sound of my childhood, it was just always there, always chirruping away in the background." Charlie cannot identify precisely why he likes sparrow chirruping, but he suspects that it is likely to stem from its ubiquitous presence when he was a child. What is notable here, perhaps, is the reference to "background." In his study of soundscapes, R. Murray Schafer divides the sounds of the environment into different categories. Birdsongs he classifies as background keynotes; sounds which may be "noticed" and "remembered with affection" when they disappear (Schafer 1977, 31, 60).

For some of interviewees, however, sparrow sounds were not just fondly remembered, but a *way of remembering*. Here's Daniel: "Yeah, I mentioned my mum sending me a DVD of a carnival in Broadbottom where I was brought up in Manchester ... and you know, the video quality was rubbish but the sound on it was actually really good – through the sort of, blurry pictures of me being dressed up as a birthday cake ... you know, through those blurry images all you could hear in the background were house sparrows, and that really took me back to the time when I was growing up, because that was the background noise, that was you know, the din of East Manchester when I grew up was house sparrows." Like Charlie, Daniel describes sparrow chirruping as "background." Yet in this story, might we not see sparrow sound as foreground rather than background? It flies from the blurry images to carry Daniel back to his childhood; as background sound on the DVD itself, it opens up a wider space of memory than that which the foreground visuals of the recorded event can provide.

Daniel was not the only interviewee to point towards this transporting quality. Maria, too, spoke about her early memories: "One of my earliest memories is of house sparrows, because on the front of the house, it was covered in really dense ivy, and there were loads of house sparrows nesting in there and my bedroom was just above some of the ivy so I used to wake up and hear the sparrows chirping and, so whenever I hear them still it kind of takes me back to my childhood." For Maria, too, chirruping sparrows were a constant feature of childhood. Accordingly, just as it does for Daniel, the sound "takes her back." Certainly, all three stories here present examples of evocative sensory memory; interviewees know what it was like to "be there" (Jones 2000). This "being there," or being *back* there, certainly seems suggestive of a rooted sense of place; the sound has firmly imprinted in memory to connect the subject to their past (Casey 2000). As Bernie Krause articulates in chapter 1, changes in the soundscapes that surround us can trigger feelings of loss, grief, and sadness, as well as disorient our sense and understanding of place. The absence of sparrow chirruping, then, may be more than just a loss of sound; it may signal the absence of (the sense of) a greater connection to a previous time or place, and the longing to return to a time of sparrow presence.

But what does it mean to be taken back? Might we think of it as something other than a simple journey back in time? If sparrow sounds take Daniel and Maria into the past, do they not simultaneously bring the past into the present?

Let us look a little more closely at how this works in Maria's narrative, for there is a subtle difference between her story and Daniel's. It rests, perhaps, in the phrase "whenever I hear them still." Where the chirruping on Daniel's DVD was produced by recorded sparrows during his childhood, Maria is trying to say that *any* sparrow sound today has this effect upon her. Sparrow sound has, for Maria, become "undone" from the present. In this sense, we suggest, sparrow sound simultaneously serves to undo the present from itself. We can say that the absence of sparrows has heightened these people's senses and produced a new way of experiencing place. This new sense of place hovers between a memory of what was and the recognition of a present that has lost something. More than that, even: their sense of "loss in place" – resonating with Albrecht's concept of solastalgia in chapter 11 – is such that were that loss to be filled, they would still sense that the present would still not be as it might otherwise have been. The present here is not a container into which the past is recalled; the sparrow sound works to *distance the present from itself.*

## London

We have seen that remembered experiences of sparrows work to (dis)place the self. Yet it is not only through individual memory that these little birds have become entwined with place. This final section finds the sparrow in "Cockney" London.[3] In doing so, it suggests not only that place may arise through the inherited memories passed down in language and stories, but that the cultural identity of Cockney London is perhaps itself spectral (Pile 2005; Sinclair 2006).

This is how Maria describes witnessing a small group of sparrows on a Covent Garden rooftop: "it was genuinely exciting, because it just felt like, you know, like proper London, this is how it was, and how it should be." Maria delights in seeing sparrows in central London. Her joy corresponds to the birds being *in place*; it gives her a sense of experiencing "authentic" London (see Ackroyd 2001, 415). Maria's excitement here might be understood as a manifestation of *jouissance* – a term from Jacques Lacan that Julia Kristeva uses to describe "the pleasure experienced in the presence of meaning" (Lorimer 2007, 922). What is important here, though, is that Maria herself is not from London, nor did she speak of ever experiencing a London filled with sparrows. Maria's sense of proper London corresponds not to "how it used to be" for her, but to how she imagines it used to be. What Maria's anecdote tells us, then, is that "we conceive of places not only as we ourselves see them, but also as we have heard and read about them" (Lowenthal 1975, 6). Her sense of place arises through a memory of inheritance, from associations made through narrative.

As conversations drew to a close, interviewees were asked what they thought would be lost if sparrows continued to decline. Elisabeth responded by saying that "Well, they figure quite a lot in our culture, I mean there's Cockney London, and they're mentioned there in connection with Cockney people who were born within a certain distance of Bow Bells or something in the east end of London, Cockney sparrows, they call them Cockney sparrows, and that would go, people wouldn't be able to relate to and understand what that meant." So Elisabeth suggests that something will indeed be lost. Yet she also points out that the term "Cockney sparrow" will remain for awhile as a tangible link to that which has been lost. Her concern here seems to be that the Cockney sparrow will remain present in language, for awhile anyway, despite the absence of real sparrows on London soil. The sparrow will linger as a ghostly linguistic presence, Elisabeth appears to be suggesting. Yet Elisabeth's use of the phrase is ambiguous. Cockney sparrow might also refer to a *person* – as in "hello me old cock sparrow," a Cockney greeting. Even without establishing to what exactly she is referring, however, we might uncover something interesting. What Elisabeth wants to say is that if sparrows are lost from London, the term Cockney sparrow – whether referring to birds or people – will be released from any visible ecological mooring, left to free-float mysteriously. What this is pointing towards, perhaps, is a longing not just for sparrows to be in London, but for Cockney London – or place – to be in the sparrows.

Yet the Cockney identity itself resists easy definition. Indeed, a stable cultural identity of place is something to which we often cling, though it remains fleeting and cannot be reduced to some "thing" (Nyoongah

2002; Till 2010). In the light of this, we might come to see that Cockney London – this imagined place to which both Elisabeth and the RSPB make reference – is itself spectral. In referencing the Cockney mythology, the RSPB is tapping into a ghost story that already haunts the popular imagination. Yet there is something else here. As with Maria's Covent Garden story, what is implied is that the original meaning of Cockney sparrow is still tangibly out there somewhere, waiting to be seen and experienced, just as Maria experienced "proper" London. In conducting the Cockney Sparrow Count, then, the RSPB is doing more than attempting to make the absent sparrow "present" (Hinchliffe 2008). What is suggested is that Cockney identity itself is still perhaps *materially* present: it *is* the sparrow that might be lurking in one's own back garden, and the sparrow *is* Cockney. It is in this sense, then, that spectral stories of cultural identity might come into play in the work of conservation. They are also, perhaps, already there: they creep into the very narratives of those involved and emotionally invested in the absence of sparrows.

## Conclusion

We began this chapter musing upon the connection between birds and place. How might the presence or absence of a particular bird shape human conceptions of place and landscape? How might conservation negotiate notions of absence, presence, and place? How might we talk about that which is absent? It was with such thoughts in mind that this study turned its attention to the decline and conservation of the house sparrow in London and the experiences of those involved in it. Various stories have been explored to suggest ways in which places – and sparrows themselves – are haunted and haunting. This haunting effect arises in a variety of ways: through the work of memory and its unsettling effects upon self and time, through memories at second hand, through the representational public aspects of conservation and perhaps even through being involved in the monitoring and recording work of conservation itself. Though it cannot hope to have spoken fully on the behalf of others, what this spectral account has hoped to point towards is an understanding of the near extinction of the house sparrow in central London as something more than a straightforward absence of a particular bird from one place. To reïterate, conservation in this case was not about preventing loss. The house sparrows project was more a response to loss in process, to the threat of disappearance, which worked to create a new, haunted landscape.

Where, then, is the sparrow? The sparrow, even though it is being lost across London, paradoxically becomes more present through a heightened awareness of its absence. Or rather, better to say that the sparrow becomes differently present: possibly present, flitting up through memory, or lurking silently in hedges, or glimpsed in signs in meadows, and in the sparrows of memory and stories. In this chapter sparrows have flitted at the edges of our writing, our sense, and our attention. Not only this, but haunting the chapter has been the fact that sparrows aren't just spectres: they are little birds that fly and eat and breed and shit. It is only through the very conjuration of absence – of a creature that is missing or missed – that presence might come into view.

## Acknowledgments

We would like thank everyone who took part in the initial research project that has culminated in the writing of this chapter: RSPB staff, volunteers, and independent professionals. Thanks are also due to Denis Summers-Smith for his hospitality and for sharing his knowledge of (all species of) sparrows. Finally, Sally Whale and Jack McCarthy must also be thanked for proofreading and providing invaluable feedback on early drafts.

### NOTES

1 Indeed, the subject of nature mourning has its own postcolonial geography (Rose and Doreen 2011). That this volume emerges out of a settler colony is perhaps not incidental, but is instead a working out of the centuries-long work of mourning for lost ways of being with non-humans.

2 The text on the sign reads: "London House Sparrows Parks Project. London's House Sparrow Population has been shrinking. This section of the park is part of a London-wide project on house sparrows. Working with the Royal Society for the Protection of Birds, Islington Council has sown different grasses and wildflowers to find out which best support house sparrows.

We will be monitoring this site for wildlife and to see how often sparrows visit."

3 London biographer and historian Peter Ackroyd describes the Cockney as "that chirpy and resourceful stereotype ... once the native of all London but in the late nineteenth and twentieth centuries identified more and more closely with the East End" (2001, 682).

## REFERENCES


Ackroyd, Peter. 2000. *London: The Biography*. London: Chatto and Windus.

Adams, Douglas, and Mark Carwardine. 1991. *Last Chance To See*. New York: Harmony.

Butler, Toby. 2006. "A Walk of Art: The Potential of the Sound Walk as Practice in Cultural Geography." *Social and Cultural Geography* 7, no. 6: 889–908.

Cammack, Paul, Ian Convery, and Heather Prince. 2011. "Gardens and Birdwatching: Recreation, Environmental Management and Human-Nature Interaction in an Everyday Location." *Area* 43, no. 3: 314–19.

Candea, Matei. 2010. "'I Fell in Love with Carlos the Meerkat': Engagement and Detachment in Human–Animal Relations." *American Ethnologist* 37: 241–58.

Casey, Edward S. 2000. *Remembering: A Phenomenological Study*. Bloomington and Indianapolis: Indiana University Press.

Challenger, Melanie. 2011. *On Extinction*. London: Granta.

Cosgrove, Denis, and Stephen Daniels. 1988. *The Iconography of Landscape: Essays on the Symbolic Representation, Design, and Use of Past Environments*. Cambridge and New York: Cambridge University Press.

Cresswell, Tim. 2004. *Place: A Short Introduction*. Oxford: Blackwell.

Davis, Colin. 2005. "État Présent: Hauntology, Spectres and Phantoms." *French Studies* 59, no. 3: 373–9.

Derrida, Jacques. 1994. *Specters of Marx: The State of the Debt, the Work of Mourning, and the New International*. London: Routledge.

DeSilvey, Caitlin. 2007. "Salvage Memory: Constellating Material Histories on a Hardscrabble Homestead." *Cultural Geographies* 14: 401–24.

Dodgshon, Robert. 2008. "Geography's Place in Time." *Geografiska Annaler* B: 901–15.

Edensor, Tim. 2005. *Industrial Ruins: Space, Aesthetics and Materiality*. Oxford: Berg.

Elder, Charlie. 2009. *While Flocks Last*. London: Corgi.

Feld, Steven. 2005. "Places Sensed, Senses Placed: Towards a Sensuous Epistemology of Environments." In *Empire of the Senses: The Sensual Culture Reader*, edited by David Howes, 179–91. New York: Berg.

Feld, Steven, and Keith Basso. 1996. *Senses of Place*. Santa Fe: School of American Research Press.

Ginn, Franklin. 2013a. "Death, Absence and Afterlife in the Garden." *Cultural Geographies.* doi: 10.1177/1474474013483220.

– 2013b. "Sticky Lives: Slugs, Detachment and More-than-Human Ethics in the Garden." *Transactions of the Institute of British Geographers*. doi: 10.1111/tran.12043.

Guardian. 2012. "The Soala in Pictures." *The Guardian*, 21 May. http://www.guardian.co.uk/environment/gallery/2012/may/21/saola-asian-unicorn-in-pictures. Accessed 21 May 2012.

Heise, Ursula. 2010. "Lost Dogs, Last Birds, and Listed Species: Cultures of Extinction." *Configurations* 18, nos 1–2: 49–72.

Hinchliffe, Steve. 2008. "Reconstituting Nature Conservation: Towards a Careful Political Ecology." *Geoforum* 39: 88–97.

Hume, Julian, and Michael Walters. 2012. *Extinct Birds*. London: Poyser.

Ingold, Tim. 2000. *The Perception of the Environment: Essays in Livelihood, Dwelling and Skill*. London: Routledge.

Jamie, Kathleen. 2005. *Findings*. London: Sort Of.

Kennedy, David. 2007. *Elegy*. London: Routledge.

Kusenbach, Margarete. 2003. "Street Phenomenology: The Go-Along as Ethnographic Research Tool." *Ethnography* 4, no. 3: 455–85.

Laurier, Eric, and Hayden Lorimer. 2012. "Other Ways: Landscapes of Commuting." *Landscape Research* 37: 207–24.

Leopold, Aldo. 1949. *A Sand County Almanac*. Oxford: Oxford University Press.

Lorimer, Jamie. 2007. "Nonhuman Charisma." *Environment and Planning D: Society and Space* 25: 911–32.

– 2008. "Counting Corncrakes: The Affective Science of the UK Corncrake Census." *Social Studies of Science* 38, no. 3: 377–405.

Lowenthal, David. 1975. "Past Time, Present Place: Landscape and Memory." *Geographical Review* 65, no. 1: 1–36.

MacDonald, Fraser. 2013. "The Ruins of Erskine Beveridge." *Transactions of the Institute of British Geographers*. doi: 10.1111/tran.12042.

Maddern, Jo Frances, and Peter Adey. 2008. "Editorial: Spectro-Geographies." *Cultural Geographies* 15: 29–295.

Merleau-Ponty, Maurice. 1962. *Phenomenology of Perception.* London and New York: Routledge.

Nyoongah, Mudrooroo. 2002. "The Spectral Homeland." *Southerley* 62: 25–36.

Pile, Steve. 2005. *Real Cities: Modernity, Space and the Phantasmagorias of City Life.* London: Sage.

Price, Jennifer. 1999. *Flight Maps: Adventures with Nature in Modern America*. New York: Basic Books.

Rose, Deborah Bird, and Thom van Dooren. 2011. "Editorial: Unloved Others: Death of the Disregarded in the Time of Extinctions." *Australian Humanities Review* 50: 1–4.

Rose, Mitch, and John Wylie. 2006. "Animating Landscape." *Environment and Planning D: Society and Space* 24: 475–9.

Rosenberg, Elissa. 2007. "The Geography of Memory: Walking as Remembrance." *The Hedgehog Review* 9, no. 2: 54–67.

Royle, Nicholas. 2003. *Jacques Derrida*. London: Routledge.

Royal Society for the Protection of Birds. 2009. *Birds of Conservation Concern 3: The Population Status of Birds in the United Kingdom, Channel Islands and the Isle of Man*. http://www.rspb.org.uk/Images/BoCC_tcm9-217852.pdf. Accessed 25 July 2012.

– 2012. *London House Sparrow Parks Project Results*. RSPB Internal Report: Unpublished.

Ryan, John C. 2009. "Why Do Extinctions Matter? Mourning the Loss of Indigenous Flora in the Southwest of Western Australia." *Philament* 15: 51–80.

Schafer, R. Murray. 1977. *The Tuning of the World*. New York: Random House.

Shaw, Lorna M., Dan Chamberlain, and Matthew Evans. 2008. "The House Sparrow *Passer domesticus* in Urban Areas: Reviewing a Possible Link between Post-decline Distribution and Human Socio-Economic Status." *Journal of Ornithology* 149: 293–9.

Sinclair, Iain. 2002. *London Orbital*. London: Penguin.

Sinclair, Iain, ed. 2006. *London: City of Disappearances*. London: Hamish Hamilton.

Summers-Smith, Denis J. 2003. "Decline of the House Sparrow: A Review." *British Birds* 96: 439–46.

Thrift, Nigel. 1999. "Steps to an Ecology of Place." In *Human Geography Today,* edited by Doreen Massey, John Allen, and Phil Sarre, 295–323. Cambridge: Polity Press.

Till, Karen, ed. 2010. *Mapping Spectral Traces*. Blacksburg: Virginia Tech College of Architecture and Urban Studies.

Tuan, Yi-Fu. 2004. "Sense of Place: Its Relationship to Self and Time." In *Reanimating Places: A Geography of Rhythms*, edited by Tom Mels, 45–56. Aldershot: Ashgate.

Tuffrey, Laurie. 2012. "Sparrow Decline Linked to Urban Noise Levels." *The Guardian*, 11 July. http://www.guardian.co.uk/environment/2012/jul/11/sparrows-urban-noise. Accessed 11 July 2012.

Yusoff, Kathryn. 2011. "Aesthetics of Loss: Biodiversity, Banal Violence and Biotic Subjects." *Transactions of the Institute of British Geographers* 37: 578–92.

Waitt Gordon, Nicholas Gill, and Lesley Head. 2008. "Walking Practice and Suburban Nature-Talk." *Social and Cultural Geography* 10: 41–60.

Wylie, John. 2007. "The Spectral Geographies of W.G. Sebald." *Cultural Geographies* 14: 171–88.

– 2009. "Landscape, Absence and the Geographies of Love." *Transactions of the Institute of British Geographers* 34, no. 3: 275–89.

# 5 Where Have All the Boronia Gone?

## *A Posthumanist Model of Environmental Mourning*

JOHN CHARLES RYAN

Loss and hopelessness. This is going to go on and on. We're never going to see these wildflowers again. What I experienced out there is gone. There's still some left, but it's under pressure all the time. All the time. It's sad for me to see this happen.

KIM FLETCHER, PERTH, WESTERN AUSTRALIA, APRIL 2013[1]

### Introduction: The Very Breath of Spring

Scented or brown boronia (*Boronia megastigma*) is a slender shrub endemic to the South-West corner of Western Australia. Said to possess a "heady, sentimental perfume" (Parker 1962, 4), the fragrant blossom is found in the heath lands and eucalypt forests between Busselton and Albany, south of the capital city Perth. Bearing small brown and yellow flowers toward the end of winter (late July–September in Western Australia), boronia was collected in the wild, shipped by train, and sold as an ornamental by Perth streetsellers in the early to mid-1900s. In 1947, novelist and columnist James Pollard (1900–1971) wrote of boronia in the Perth newspaper *The West Australian*. Evoking his experience of the wildflower in sensuous terms, Pollard extols boronia's "perfume stirring memories" (4). He shares a "scented memory" with an onlooker in the street as Pollard – then in his middle years – recollects picking boronia in his youth. In Pollard's account, the flower's aesthetic appeal and the regional economic network, to which it was integral, bridged the divide between city and country: "The young people of the boronia country go out in the rainy dawns of July and August, to gather and dispatch to the city the flower that in this season gives to Perth one its few town cries" (Pollard 1947, 4).

Boronia's "sentimental" olfactory signature sustained cultural memory and identity in urban Perth and in the South-West region for as Parker (1962) commented, "brown boronia is a West Australian institution – the very breath of spring" (4). With boronia flowers under increasing demand, regulations were implemented during the mid-1900s to control the harvesting of wild plants and the production of fragrant distillations known as "otto of boronia" (*Western Mail* 1936). However, by the 1960s, concerns about the decimation of boronia began to intensify in popular articles attributing the decline of the species to overharvesting, habitat destruction, bushfires, and the replacement of wild collecting (and the city-country ecocultural networks it fostered) with the promotion of cultivated varieties (Parker 1962, 4). During this time, the viability of the species in the wild became jeopardized. While boronia populations now appear stable (Packowska 2013), there are nevertheless many cultural losses linked to the decline of the species as a non-human presence embedded in a network of human-non-human relations. The trains no longer effuse boronia perfume. The streetsellers no longer intone the call: "Fresh, sweet-scented boronia! Sixpence a bunch, bor-o-nee-a!" (Parker 1962, 4). The blossoming of boronia – heralding seasonal transition – no longer signifies the rising "breath of spring."

As a result, the cleft between urban and rural life in Western Australia – once traversed sensuously by the "sentimental" perfume of boronia wafting along rail lines – widens with loss. The rapid urban development of Perth has led to encounters with ersatz boronia where, for example, an online search for the plant genus results in an odd panoply of hits: a Department of Corrective Services facility, an online clothing outlet, an investment company, and a web content management system. Once a "perfume stirring memories," boronia sadly has become a figment of Perth's urban imagination – an artifact confined to the recollections of a generation for whom the wildflower was an "institution" in itself.

## Perth, South-West Australia: A Case Study in Botanical Loss

Anstey-Keane Damplands was totally chained by the City of Armadale, which wanted to put a big recreational park there about 40 years ago. Fortunately it was only chained and left undeveloped, so the bush recovered quite quickly after being cleared. But that's why you don't see many old trees. They're all regrowth. Believe me, there were lots of big trees. They won't be seen again in our lifetime because you need 500 years to grow big zamia palms or big banksias.

DAVID JAMES, FORRESTDALE, WESTERN AUSTRALIA, SEPTEMBER 2009[2]

With pangs of melancholic nostalgia, I use boronia as a synecdoche and entry point into the broader investigation of the continuing loss of flora in South-Western Australia, the focus of this chapter. The South-West region refers to the lower corner of the state of Western Australia from Shark Bay in the upper northwest to Israelite Bay east of Esperance in the southeast, and including the metropolitan Perth area. The biodiversity of the South-West – particularly its flora – is of international renown; the region is one of the most botanically diverse Mediterranean ecosystems in the world. A global biodiversity epicentre, the South-West exhibits remarkable biological endemism (Hopper 1998, 2004) and is the only internationally recognized Australian "hotspot" (Conservation International 2008, 2013). Approximately 35 per cent of the region's plants are endemic; that is, found to occur in uncultivated conditions only within its biogeographical boundaries. In the late 1800s, German botanist Baron von Müeller applied the term "botanical province" to the region, observing its unique floristic communities and unusually high rates of endemism (Beard 1979, 107–21).

As the primary urban area within the South-West, Perth is also a biodiverse yet rapidly-changing city. The prominent Australian botanist Stephen Hopper affirms the "tremendous diversity" exhibited by the South-West, distinguishing Perth as "one of the world's most biodiverse cities, especially in relation to plants" (Perth Biodiversity Project n.d., 1). However, the increasing loss of remnant bushland threatens Perth's botanical conservation and compromises the long-term viability of its flora. Indeed, the same factors threatening Perth's unique flora also threaten the plant life of urban areas globally (Farr 2012; Hostetler 2012; Knapp 2010). More specifically, in Australian cities, plant diversity is impacted by a variety of environmental and anthropogenic pressures, including but not limited to bushland development (McKinney 2005), plant diseases (Shearer et al. 2007), dry land salinity (Albrecht 2005, 53; Beresford et al. 2001), and demographic shifts typical of many urban areas (Girardet 2008). Considered one of the world's most ecologically diverse cities,

Perth faces the imminent possibility of losing its ecocultural uniqueness, especially in the swiftly expanding southern suburbs between Armadale and Fremantle (Giblett 2006; Giblett and James 2009). While adversely affecting the functionality of ecosystems, these pressures also diminish meaningful human exposure to the natural world and impede intimate interaction – or what Donna Haraway (2008) refers to as co-constitutive entanglement or "becoming with" – between humans and non-humans (4, 88).

Whereas conservation biology and other scientific disciplines address the loss of biodiversity from specialist and technical perspectives, human experience – memory, emotion, sensoriality – bears discernible absences that otherwise go unrecognized and under-researched (Ryan 2012, chapter 1). How can we mourn the increasing loss of plant species and their habitats in a quickly transforming urban environment like Perth? Which philosophical and conceptual orientations can guide, nurture, and sustain our processes of mourning environmental loss – on both an individual and collective basis?

In this chapter, I outline a model of environmental mourning based in connectivity and interdependence – qualities intrinsic to the science of ecology that convinces us, with ever-increasing certitude, of the relational lives of organisms (see, for example, Smith and Smith 2012). Through a case study of Perth's flora and its human advocates, I propound a posthumanist theory of environmental mourning – one that relies on notable developments in multispecies theory and the ecological humanities. A posthumanist theory extends recent work, conducted from different disciplinary perspectives, by Glenn Albrecht (2005, 2010, Albrecht et al. 2007), Donna Haraway (2008), Timothy Morton (2010), Deborah Bird Rose (Rose and Robin 2004; Rose 2008, 2011a, b), Thom van Dooren (2010, 2011), Cary Wolfe (2010), and Kathryn Yusoff (2011), as well as previous research into environmental mourning and the cultural significance of plants in the South-West region (Ryan 2012, chapter 11; Trigger and Mulcock 2005).

Forwarding a non-binary position between the human, "unhuman, nonhuman, or 'more-than-human,' or possibly even inhuman" (Morton 2010, 252), a posthumanist model of environmental mourning will be developed in respect to the loss of botanical habitats in Perth and the impact of this loss on the sense of place of my respondents. The ethnographic material that informs this chapter represents a small part of a series of interviews performed since 2009 with South-West Australians in order to understand their commitments to botanical conservation in the context of Perth's declining flora. What started as a one-dimensional interest in my interviewees' aesthetic attitudes toward South-West plants – their beauty, sublimity, picturesqueness, or ugliness and the relationship between environmental aesthetics and community appreciation of the region's biodiversity – evolved into a multi-nodal study of human sensoriality and flora. During the course of my research, memory and mourning have been implicit formations, generated at the intersection of scientific knowledge, sensory experience, and emotional recollection. Throughout this process I have found it impossible to disentangle memory and mourning from the understanding of how one's sense of place develops in relation to scientific and non-scientific understandings of plants – a broader phenomenon termed "floratopaesthesia" or literally the sense of place fostered through sensory experience of flora (Ryan 2012, chapter 13).

## The Humanist Aspects of Freudian Mourning

Some of the orchids, like the spider orchids, are not as plentiful, especially the white spider orchids. Weeds smother small plants. Dodder laurel creeper climbs on paperbarks. It's still here. Many local swamps have been bulldozed in the last thirty years. Weeds have changed the appearance of the lake forever. I don't know what the answer is.

DAVID JAMES QUOTED IN GIBLETT 2006, 97

A multispecies theory of environmental mourning eschews the problematically individuated and human ego-focused principles of Freudian mourning. A theory of environmental mourning should transcend the individuated subjectivity of Freudian mourning that positions an object of loss – a habitat, a plant, an animal – in a dichotomizing relationship to the grieving ego of the subject – a human mourner, a conservationist, an interviewee. I begin this section by critiquing the insufficiency of Freudian mourning to account for the loss of the networks – interdependencies, connectivities, relationships – between living creatures and their living and non-living milieux. Freudian mourning foregrounds human subjectivity and, consequently, backgrounds the ecocultural interconnections and multiple subjectivities with which a human subject perpetually transacts and intersects.

Moreover, Freudian mourning propounds an anthropocentric mode of responding to loss. I critique the inherent anthropocentrism of Freud's theory of mourning and melancholia (Clewell 2002; Dodds 2011; Homans 1989), particularly in relation to environmental grief and "ecomelancholia" (James 2011). A theory of environmental mourning should account for the loss – of the nodes and non-human "things" with which a human subject has established ecological intimacy – in ways that inherently counter tendencies toward human narcissism throughout the different manifestations of environmental mourning. In developing this critique, I deviate from a body of scholarship that applies psychoanalysis constructively to the environment and human-nature relations (for example, Ley 2008). Freudian mourning centralizes the human through a framework of hyper-individuated subjectivity. Accordingly, I shall briefly explore the connections between humanism and Freudian psychoanalysis (for example, Dodds 2011). Humanism refers broadly to an aggregate of philosophies prioritizing human agency, empiricism, and rationalism over faith-based meaning-making.

As Fowler (1999) points out, "humanism is concerned with the secular, rather than the religious; with this life, as opposed to projections of life beyond death when the *human* no longer exists; and with the immediacy of the temporality of human existence rather than any suggestions of the eternal nature of the human" (9, emphasis in original). For Fowler (1999), humanism reached its apotheosis in the seventeenth and eighteenth century Enlightenment ideals of Imannuel Kant and Isaac Newton (16). Hence in learning about and interacting with the natural world humanism emphasizes "empirical science and critical reason" (Wolfe 2010, xiv) over the supernatural or divine. In his critique of humanism, Cary Wolfe contends that the notion of "the human" is constructed in opposition to animality in order to liberate the human from biological and evolutionary realities – and to transcend the immanence of embodiment (ibid., xv). Humanism pivots on "normative subjectivity – a specific concept of the human" (ibid., xvii), which polarizes the human and the animal and, consequently, hyper-separates humanness from vegetality.

While an extensive treatment of Freudian theory – for example his writing on grief in "On Transience" (1916) and *The Ego and the Id* (1923) – is out of the scope of this chapter, I will identify some of the humanistic tendencies of Freud's theory of mourning in his 1917 essay "Mourning and Melancholia." My critical position is that Freud's distinction between "conscious" mourning and "unconscious" melancholia provides a conflicted basis for a theory of environmental mourning accounting for connectivity and interdependence between species – one that circumvents the pitfalls of humanistic subjectivity and egocentrism that multispecies theory and the ecological humanities seek to redress. Early in "Mourning and Melancholia," Freud defines his mourning framework as comprising both tangible and abstract "objects" or as "regularly the reaction to the loss of a loved person, or to the loss of some abstraction which has taken the place of one, such as one's country, liberty, an ideal, and so on" (1968, 243). Whereas mourning is a normal grieving process in response to loss, after the completion of which "the ego becomes free and uninhibited again" (ibid., 245), melancholia is a pathological outcome. As an unconscious response to loss, melancholia entails "profoundly painful dejection, cessation of interest in the outside world, loss of the capacity to love, inhibition of all activity, and a lowering of the self-regarding feelings" (ibid., 244).

Freud's theory of mourning pivots on a binary between mourning as the extension of grief at the loss of the world and melancholia as an inward projection of grieving emotions linked to human narcissism: "In mourning it is the world which has become poor and empty; in melancholia it is the ego itself" (ibid., 246). For Freud, the process of mourning oscillates between the mourning subject and the mourned object, whereas melancholia inverts grieving emotions into a state of solipsistic self-reflection in which the world and human-non-human interconnectedness vanish. Cultural studies scholar Rod Giblett (1996) rephrases the Freudian distinction as "in mourning the world is experienced as loss whereas in melancholia the ego is experienced as loss" (177). Significantly, Freudian mourning posits the environment as an object of a human subject's mourning – a backdrop to ego-driven loss that, when pathological, instigates melancholia. Freud articulates the human grief process through which the mourner severs emotional ties to the lost object, entailing a "detachment of libido" (Clewell 2002, 44) marking the completion of mourning. In the final analysis, mourning ceases when the mourner dissolves emotional linkages to the lost object-cathexis. The severance liberates the human ego to focus on a new object, creating a substitution for the absence of the lost one – a process that seems to be driven by narcissistic self-identification rather than an ecological sense of interdependence.

The dyad of Freudian mourning and melancholia circumscribes the traditionally unified subject in which human allegiance is to personal desire rather than to the "objects" of desire (Clewell 2002, 1) and even less so to the communities or networks of desire, affection, and love (Rose 2008, 2011a). For the mourning of plants, what is lacking in Freudian thought is an expression of human interrelationship to the world, beyond the pathologies of a strongly individuated subject entrapped in the (quite possibly self-perpetuating) cycle of melancholia. Clewell (2002) argues that "Freud's mourning theory has been criticized for assuming a model of subjectivity based on a strongly bounded form of individuation" (1). In the context of Freudian mourning, the threatened, endangered, or extinct plant would be mourned not for the loss of a being and its living embeddedness, but rather as a reflection of the subject's inevitable decline. Strikingly anthropocentric, Freud's "bounded form of individuation" requires object-subject dynamics to produce transferences between the mourning subject and the mourned object. In light of this, I argue that Freudian mourning constrains the emergence of environmental mourning based in connectivity and interdependence.

In comparable terms, environmental psychologist Joseph Dodds (2011) asserts that Freudian psychoanalysis embeds an objectivist worldview stemming from the Enlightenment ideal of emotionally detached rational inquiry. As Dodds argues, quoting *The Future of an Illusion* (1927), Freud's theory of mourning sets the imperatives of psychoanalysis in opposition to the drives of nature: "The principle task of civilization, its actual *raison d'être*, is to defend us against nature. We all know that in many ways civilization does this fairly well already, and clearly as time goes on it will do it much better" (Freud 2008, 115). As Dodds goes on to explain, the Freudian dichotomies – between masculinity and femininity, civilization and nature, order and chaos, and autonomy and dependency – are definitive tropes found in the thinking of the Enlightenment, modernity, and patriarchy. Freudian psychoanalysis furthermore carries elements of the Enlightenment project to bring unconsciousness to consciousness, lending greater weight to the rational ego and resulting in a "master-slave system of absolute binaries, and an attempt to maintain an illusory autonomy and control in the face of chaos" (Dodds 2011, 32).

As a counterweight to Freudian mourning and its humanistic inclinations, "multi-species mourning" will be theorized next in relation to the loss of Perth's flora through theorists associated with three domains: posthumanism, the ecological humanities, and multispecies theory. While examining how concepts of environmental mourning have hitherto been developed by environmental scholars, independently of humanistic traditions, I will argue for the applicability of these three paradigms – located in literature, philosophy, cultural studies, and animal studies – to an emergent perspective on flora called "human-plant studies" (Ryan 2013, chapter 6), which most notably integrates emerging notions of plant subjectivity into critical humanities-based botanical research (Marder 2013).

## Toward a Posthumanist Model of Environmental Mourning

> Wreath flowers (*Leschenaultia macrantha*) were rare. If you picked those, by the time you stood up, it started to wilt. An amazing flower – just pick it up and it's gone. They were up around Mullewa and Bullardoo Station. There's a big hill there. It was iron ore and, sadly, they loaded the hill onto trucks and railway cars and carted it away. Tallering Peak was a really important place.
>
> NOEL NANNUP, MOUNT LAWLEY, WESTERN AUSTRALIA, JULY 2010[3]

Rather than Freudian subject-centred mourning, a posthumanist framework relates the mourning of declining plants to bodily experience and sense of place while counterbalancing anthropocentrism through the decentring of human subjectivity. Indeed, as one of its major aims, posthumanism negotiates the "decentering of the human" (Wolfe 2010, xv) in the construction of knowledge and meaning. Recent posthumanist developments in multispecies theory – most prominently the notion of "companion species" (Haraway 2008), and in the ecological humanities specifically the concept of "connectivity ontology" (Rose and Robin 2004) – offer posthumanist perspectives for theorizing the mourning of vanishing non-humans. Moreover, in decentring human subjectivity without nullifying the essential role of experience, a productive direction proffered by multispecies theory is a focus on human-non-human sensory entanglements. Barad (2010) stresses, "entanglements are not a name for the interconnectedness of all being as one, but rather specific material relations of the ongoing differentiating of the world. Entanglements are relations of obligation – being bound to the other – enfolded traces of othering" (265). I depart slightly from Barad in her association of entanglement with material obligations only. Human-non-human entanglement also entails intangible mnemonic, emotional, and sensual interconnections that develop in conjunction with ongoing material transactions. In this context, posthumanist mourning is inclusive enough to account for both human-non-human entanglements and the plant species impacted by extinction pressures, such as habitat loss, as well as the often devastating emotional and phenomenological depletion of human sense of place that occurs.

In order to articulate more clearly a theory of environmental mourning, I will outline some of the tenets of posthumanism in order to clarify its relevance to the decline of a biodiverse world. Through the decentring of subjectivity, posthumanist environmental mourning negotiates the positioning of the human in the sensuous fabric of the world, in the materiality of an interdependent ecological existence. For Wolfe (2010), posthumanism "opposes the fantasies of disembodiment and autonomy, inherited from humanism itself" (xv). In terms of Barad's notion of entanglement, posthumanism embeds human subjectivity in the material demands of place through embodied experience and awareness of tacit human-non-human transactions and interconnections.

While the term "posthumanism" emerged in the mid-1990s in the humanities and social sciences, a notable antecedent is systems theory, developed by Gregory Bateson, Norbert Wiener, and others during the 1950s (ibid., xii). Early developments in system theory attempted to dislodge human subjectivity "from any particularly privileged position in relation to matters of meaning, information, and cognition" (ibid.). But where does environmental mourning lie within Wolfe's interpretation of posthumanism, which is largely contextualized through human-animal studies? The first motion toward a posthumanist model of botanical mourning, therefore, is to interface Wolfe's analysis of the decentring of the human with philosophical (Marder 2013), literary (Ryan 2012), and scientific (Trewavas 2002, 2003, 2006) developments in plant research, in order to reach a theory of botanical mourning applied to interpret some of my interviewees' comments in the following section.

In addition to its focus on relationality, posthumanist mourning is *specific* (multispecies or interspecies); the mourning of botanical loss tends to involve intimate knowledge of actual *species*. Acknowledging the layers of ecocultural loss embedded within human-non-human entanglements, posthumanist mourning provides scope for the mourning of individual non-humans in order "not to reduce them to interchangeable cogs in an ecosystem machine" (van Dooren 2010, 273). Indeed, as the following section will demonstrate, the contraction of plant diversity underlies the impoverishment of my interviewees' sensory lifeworlds, memory networks, and emotional well-being. While posthumanist mourning will always remain *human* – after all, we are human beings who mourn and whose affective relations contract when other species are lost – it can engender within us an awareness of Freudian mourning's dangerous hyper-subjectivity. The decentring of the human is not simply a turn from anthropocentrism to ecocentrism (from culture to nature), but rather a return to a sense of relationality between species as the essence of mourning.

This is not to say that humans are the only beings who mourn: other mammals do (King 2013), and we should not dismiss the possibility of emotions and mourning in plants. Although important to take into account and explore further as part of a posthumanist theory, the mourning experienced by plants is unfortunately out of the scope of this chapter. The potential of posthumanist environmental mourning, therefore, is its capacity to reflect the "mosaic quality that recognises causality between the self, the other and the environment" (Ryan 2012, 278) – its appreciation that environmental loss fragments multispecies entanglements and renders human experience of place monochrome and *life-less* (literally). The crucial interspecies dimension of environmental mourning, furthermore, encompasses the roles of exotic or invasive flora – exemplified by the intrusion of yangee rushes into endangered Forrestdale Lake wetland ecosystems south of Perth (Giblett 2006). Interspecies mourning comes to reflect the dynamics of endangered places, including, in the case of Forrestdale, the colonization of a wetland environment by an aggressive plant, the ensuing displacement of indigenous biota and the decline of the aquatic ecosystem over time. In many examples of environmental loss, the human or non-human "aggressor" (i.e. government agency, corporate developer, introduced animal, or exotic weed), such as the yangee rush, becomes an object of anger and despair as much as the lost species becomes an object of mourning and grief. A posthumanist model suggests that the understanding and experience of mourning can become intrinsically ecological, moving from the disappearance of individual species or the impoverishment of the human life-world towards an ecology of mourning that is much more complex, inclusive, intersubjective, evocative of place, and deserving of research.

In terms of its *specificity*, the title of this chapter – "Where Have All the Boronia Gone?" – responds in part to Timothy Morton's work on elegy and his development of "dark ecology" with respect to environmental loss. In the section "Eco-elegy, or Where (Will) Have All the Flowers Gone?," Morton (2010) propounds an essentially negative view of environmental mourning, claiming that "we cannot mourn for the environment because we are so deeply attached to it – we *are* it" (253, emphasis in original). In other terms, Morton (2010) contends – in response to Judith Butler's analysis of melancholia – that "the truest ecological human is a melancholy dualist, mourning for something we never lost because we never had it, because we *are* it" (253, emphasis in original). Through statements such as these, Morton firstly deconstructs "environment" as a signifier, secondly suggests the impossibility of environmental mourning, and thirdly attempts to counter the tendency in nature writing to construct the natural world as a backdrop for human grief and – tangentially – as an object-cathexis of human mourning in Freudian terms. Importantly Morton (2010) asks "what happens when this backdrop becomes the foreground?" (253). Providing no firm alternative of his own, Morton characterizes environmental mourning in its current state (Freudian) as a "narcissistic panic [that] fails fully to account for the actual loss of actually existing species and environments" (ibid., 255). Although deconstructive and ultimately nihilistic, failing to offer conceptual alternatives to what he brands as Freudian narcissim, Morton's critique does invigorate the need for a theory of environmental mourning drawn from multispecies thinking – one that goes beyond the generalizability of the term "environment" and alternately accounts for "actually existing species" in all their specific materialities and sensuous expressions. In eliding firm distinctions between human and non-human subjectivities – "we *are* it" – Morton calls for a new theory of environmental mourning, one reflecting the philosophy of ecology as "thinking how all beings are interconnected, in as deep a way as possible" (ibid.).

The posthumanist approach I propose also looks toward constructive, community-based, and human-focused interpretations of environmental mourning, exemplified by Glenn Albrecht's (2005) research into solastalgia or "the pain or sickness caused by the loss or lack of solace and the sense of isolation connected to the present state of one's home and territory" (48). Whereas nostalgia refers to one's sense of loss when displaced, solastalgia encompasses the feelings of mourning when one experiences firsthand the deterioration of his or her home place or bioregion and, as Albrecht explains in this collection, reflects the "deep form of existential distress when directly confronted by unwelcome change in [a] loved home environment" (chapter 11, 292). Solastalgia is the "lived experience of the loss of the present" (Albrecht 2005, 48) linked to the decline of the capacity of a place to nurture a sense of home or a feeling of love as "topophilia" (Tuan 1974) or "biophilia" (Wilson 1986). Elsewhere, Albrecht et al. (2007, S96) define solastalgia as "place-based distress in the face of the lived experience of profound environmental change." Evolving from ethnographic work in the Upper Hunter Region of New South Wales, Australia, solastalgia is the "homesickness one gets when one is still at 'home'" (Albrecht 2005, 48) and is most intense as a result of sustained sensory and emotional experience of the environment's decline over time: "The people of concern are still 'at home,' but experience a 'homesickness' similar to that caused by nostalgia. What these people lack is solace or comfort derived from their present relationship to 'home,' and so, a new form of psychoterratic illness needs to be defined ... solastalgia exists when there is the lived experience of the physical desolation of home" (Albrecht et al. 2007, S96). Indeed, the factors precipitating solastalgia can be either environmental or anthropogenic, and can include "drought, fire, and flood" as well as "land clearing, mining, rapid institutional change" (Albrecht 2005, 48) – the latter most evidently in my interviews with Perth conservationists.

While Albrecht's solastalgia accounts for the "psychoterratic" effects of generalized place-based deterioration, however, posthumanist mourning provides a conceptual perspective for addressing the decline of "actually existing species" (Morton 2010, 255) – in this instance plants. Environmental mourning accounts for the emotional bonds between people and places in response to what Porteous (1989) terms "topocide" (230) or the annihilation of place and what Giblett (1996) calls "aquaterracide" (68) or the killing of wetlands. The *specific* focus of posthumanist environmental mourning – on urban botanical loss here – foregrounds the emotional and material connections people develop to flora. Additionally, while Albrecht's solastalgia has been developed in reference to the effects of mining and drought, the loss of Perth's flora expands the contexts of despair in which human well-being is impacted adversely by environmental loss. As proposed here, a posthumanist approach to environmental mourning examines the nature of loss itself. Yusoff (2011) argues that loss "can be considered a measure of feeling that proceeds from vulnerability, from the loss of a self that is constituted in relation, rather than in isolation" (579). Following Yusoff, multispecies mourning brings attention to the constitution of human-non-human subjectivities and the relational quality of loss beyond the anthropocentric Freudian dyad of mourning and melancholia.

The model of environmental mourning that I am sketching also follows recent work by ecological humanities scholars on the social and philosophical impacts of extinction. In his research into the rapid decline of Indian vultures, Thom van Dooren (2010) summarizes poignantly the significance of a multispecies perspective: "Attentiveness to the relationality and interdependence of life is particularly important because the death, and subsequent absence of a whole species, unmakes these relationships on which life depends, often amplifying suffering and death for a whole host of others" (273). Significantly, responses to place-based deterioration often involve the scientific management of populations and species, rather than an acute awareness of the pain experienced by individual humans and non-humans. Van Dooren insists that "individuals – as ethical subjects – be brought back into a conservation discourse that is saturated with species, habitats and ecosystems" (277). A posthumanist model of environmental mourning acknowledges individuals as "relational beings" and that mourning, as a consequence, must be relational, multispecies, and multi-sensorial in order to be emplaced as a phenomenon in the world.

## Mourning the Decline of Flora in Southern Perth: Voices from the Community

What we took for granted is now threatened by housing and the expansion of agriculture and roads. The piece of bush that we took for granted as kids is now being threatened. It's a bitter shame. I spend more time trying to protect wildflowers than actually going out trying to enjoy them for what they are.

DAVID JAMES, FORRESTDALE, WESTERN AUSTRALIA, SEPTEMBER 2009

In this section, I will further advance a posthumanist model of environmental mourning through analysis of interviews with residents of the southern Perth suburbs. Since 2008 I have been conversing with botanists, conservationists, educators, wildflower enthusiasts, and others with long-standing relationships to the city's rapidly disappearing plant life. Drawing on ethnographic methods and interrogating theories of human memory, I have found that the decline of plant species comes with the loss of olfactory, gustatory, haptic, acoustic, and visual networks between humans and non-humans (Ryan 2012, chapter 8). For this discussion, I will focus in detail on two interviews with Armadale area conservationists David James and Kim Fletcher – both of whom express the complexities of environmental mourning, yet from differing perspectives. My interviewees are well-informed, passionate about, and closely involved with Perth's flora; for example, Kim Fletcher has collaborated with Western Australian scientists in the taxonomic identification of orchids. In addition to the interviews, I will refer to previously published oral histories (Giblett 2006) and, where relevant, archival and literary material, exemplified by James Pollard's boronia writings. The eclecticism of this chapter as a whole reflects an integrated approach to humanities-based plant research, examining a wide range of sources and theorized as "cultural botany" (Ryan 2012, chapter 1).

Themes of loss are evident in an interview with David James of Forrestdale, Western Australia. A passionate activist and self-trained botanist born in the early 1950s, David has lived near Forrestdale Lake all his life. Forrestdale Lake is located on the Swan Coastal Plain – shared by Perth to the north – and protects numerous indigenous floral, faunal, fungal, and insect communities (Giblett 2006). Nearby Anstey-Keane Damplands is one of the most botanically significant places on the Swan Coastal Plain and more biodiverse than popular Kings Park, adjacent to Perth Central Business District (Giblett and James 2009). Anstey-Keane lies at the northern tip of the Pinjarra Plains, a system of flat damplands – moist, shallow sinks – including the most desirable soil on the Swan Coastal Plain for pasture and development (Beard 1979, 27). A member of several local conservation organizations, David conveys his exasperation over the ostensibly insurmountable pressures on the environment posed by suburban development. For David, Forrestdale Lake Nature Reserve and Anstey-Keane Damplands constitute "emotional geographies" (Bondi, Davidson, and Smith 2005) – places linked to his extensive conservation commitments in the quickly changing southern areas of Perth. David's involvement with different organizations charged with protecting land is "a period of, shall we say, activism [with] organizations that are actually trying to preserve the environment." Environmental activists, such as David, exemplify a single-minded devotion to maintaining a community-based sense of place through the protection of plants.

During David's childhood in the 1950s, the bush around Perth seemed limitless and impervious to modern suburban expansion: "We'd walk through bush to catch the school bus." His memories suggest a popular contemporaneous perception of the flora as all encompassing and inexhaustible. The former abundance of the bush intermingles with recollections of the encroachment of development and his attendant emotions of powerlessness: "In those days, the bush was everywhere. Nowadays, you realize how threatened it is, but in those days, it was common. We'd walk through bush to catch the school bus. We took it all for granted. Even in those days, people were destroying bushland but, because there was so much, as a kid, you just accepted it." David's recollections reveal how environmental mourning – in Albrecht's terms "solastalgia" or the distress of witnessing the decline of one's home place while one is still there – can involve detailed information about a land transformed by the juggernaut of development. His childhood reverie has been displaced by an anxiety over the manifold threats to the local landscape, compelling him to align with conservation initiatives: "I spend more time trying to protect wildflowers than actually going out trying to enjoy them for what they are."

David's interview also demonstrates environmental mourning as a wellspring of emotions about the natural world. Moreover, the emotion-filled memory narratives of amateur botanists and local conservationists like David can impart a sense for the scale of change. For example, the proliferation of exotic plant species on roadsides is a distinct difference over David's lifetime: "Years ago roads were narrow and roads with vegetation in good condition were quite normal." His recollections confirm the progressive incursion of exotic plants on the west side of Forrestdale Lake and the gradual disappearance of certain orchid species. These distinctive intrusions to the composition of his home place have occurred within his memory: "Nowadays this side of the lake's pretty much weed infested, but in those days the weeds weren't quite so bad. We used to get orchids growing alongside the road here, spider orchids and different species growing amongst the weeds." Slow-growing plants, such as banksias and the zamia palms, have been severely affected by destruction of bushland areas, specifically Anstey-Keane Damplands and Forrestdale Lake (Giblett, 2006). According to his direct experience, the climax character of the land has been permanently altered, despite positive claims about the merits of bushland reconstruction by environmental restoration techniques: "After 30 or 40 years, it looks quite natural. But believe me, if you went back before that, there was a lot of big stuff in there. That won't be seen again in our lifetime because you need 500 years to grow big zamia palms or big banksias." David evokes one of the distinguishing qualities of the South-West flora: its ancientness and slow-growing habits resulting from low soil nutrients and intensive solar exposure (for example, Breeden and Breeden 2010). Certain plants require hundreds of years to reach a mature state, amplifying environmental loss in a manner reminiscent of the mourning of old-growth forests farther south in the South-West region (Trigger and Mulcock 2005).

Like David James, Kim Fletcher has lived most of his life near Forrestdale Lake Nature Reserve and Anstey-Keane Damplands, near Armadale in the southern suburbs of Perth. Kim was born in 1937 in Cottesloe, Western Australia, to a family of wildflower enthusiasts and amateur orchidologists. As an educator, in Armadale and elsewhere in the state, he developed strong teaching skills he now uses in his role as a volunteer guide with Kings Park and Botanic Garden. Some of Kim's earliest memories of Perth's wildflowers relate to orchid collecting on Sunday afternoons with his father and mother. In order to locate wildflowers,

Kim's family frequented swathes of original bushland between the railway line and the main highway to Perth where, in particular, orchids proliferated. However, as Kim explains, much of this land has now been converted to suburbia: "It was ideal for collecting orchids, particularly spider orchids [*Arachnorchis* spp. formerly *Caladenia* spp.] and enamel orchids [*Elythranthera* spp.], because it was sandy country. This, of course, has now disappeared altogether as bushland. It's covered with suburbia, with houses now. That started probably in the late 1950s and early 1960s." Kim evokes a sense of multispecies mourning in the following excerpt in which he describes locating a rare hammer orchid (*Drakaea* spp.). As an expression of posthumanist mourning, his recollection involves other species – a marri tree (*Corymbia calophylla*) and cormorants (*Phalacrocorax varius*) – as well as sensuous detail about the attractive visual qualities of the orchid. The land, converted to suburbia, has been shared amongst different species as a community; consequently Kim's manifestation of mourning exhibits relational, multispecies, and sensory qualities. However, his initial curiosity and child-like awe shift abruptly to hopelessness – hopelessness about ever finding the orchids in his home place again: "Under a marri tree, I found a heart-shaped, glossy green, beautiful leaf lying flat to the ground with a stalk coming up. I thought, 'Oh, what's this? It must be an orchid.' I visited practically every day and eventually, of course, it came out. It happened to be a hammer orchid, the first I'd ever seen. Apparently now these are declared rare flora. I did find the same orchid elsewhere over in the Forrestdale Lake area, but they were the only two I've ever found. There's no hope of finding them because the land is degraded badly and is going to be surrounded by suburbia in the near future. Incidentally, it was also a rookery for cormorants." The multispecies ambit of Kim's mourning furthermore includes references to spearwood (*Kunzea ericafolia*) and the intrusion of exotic plants, such as arum lilies (*Zantedeschia aethiopica*), leading to the formation of hybrid botanical communities intermingled with the new housing estates. Kim expresses other feelings of loss – of returning to his home ground in hope of finding jewel orchids but repeatedly being disappointed. In the following excerpt, there is also an obvious connection in Kim's memory between spearwood and jewel orchids (*Anoectochilus* spp.), indicative of the multispecies community he now mourns: "There were *Kunzea ericafolia* growing along

Forrest Road. They were so common. Of course, during the 1980s, it was cleared and houses put on it. There is still some bush in the area, but it's been invaded by arum lilies and goodness knows what. That site has gone, I'm sorry to say. Although Armadale Golf Course is contiguous to the conservation area around Forrest Road, the nature reserve and the orchids have disappeared. I've been back many times trying to find jewel orchids." Local flora, especially orchids, signify place in Kim's lived experience – reflecting the development of floratopaesthesia or sense of place through plants. Kim narrates his experience of finding ghost orchids or rattle beaks (*Lyperanthus* spp.) in Wungong Gorge east of Armadale. He later came across a surprisingly large cluster of ghost orchids in the city of Armadale, not far from Wungong. Along with the ghost orchids, Kim recalls the blue china orchid (*Caladenia gemmata*) and diverse species of banksia in the Peel Estate area outside Armadale. As an orchid lover, the ghost and blue china orchids represented peak, affective, and multispecies experiences for Kim – "it took my breath away." However, these sites of memory and meaning have been razed: "Not as far as the bush to the east of Armadale, which is forest country, I would go exploring. One of the big finds was in Wungong Gorge. I found a patch of *Lyperanthus*. It's called the ghost orchid or rattle beak, but that's long gone. However, in Fletcher Park [in the City of Armadale] I found about forty ghost orchids. I also remember getting out into the Peel Estate where there were *Caladenia gemmata* under the banksia. It is a little blue orchid and that was right through. It took my breath away when I found that. But it's all been cleared." In another excerpt, Kim's memories of rabbit or hare orchids (*Leptoceras menziesii*, formerly *Caladenia menziesii*) capture the sensuousness of the species growing together in a mass. Nevertheless, the poignancy of Kim's recollection is tinged with powerlessness and despair, marked also by an awareness of the "weeds" that have displaced the orchids: "There was a huge patch of what was formerly known as *Caladenia menziesii* or the rabbit orchid. It was a big mass. All the leaves and flowers were shimmering right there. Many times I've gone back with a tear in my eye, looking at where they were among the weeds close to the road." Solastalgia is an apposite framework to describe Kim's mourning of environmental loss. Infused with emotional and family resonances, his home place has changed drastically as the orchids he loves disappear at alarming rates:

"It's a very sad time for me when I go back to these places that I used to frequent so often as a kid when I got to love orchids." Kim's statements underscore the idea that environmental mourning is ecological – or, in Morton's terms, "interconnected, in as deep a way as possible" (2010, 255). The mourning of a plant species is rarely of that species in isolation but rather within its ecological milieu – within its multitude of relations. For Kim, the orchids of the Perth suburbs invigorate recollections of other plants, animals, and birds, as well as a sense of place and personal identity constituted over many years of living in the Armadale area. As Yusoff aptly states, loss "requires mourning and grieving for the destruction of a relation and those subjects that are constituted through that relation" (2011, 579).

Highlighting the home place shared by Kim and David, Rod Giblett's *Forrestdale: People and Place* (2006) is based on oral histories from Forrestdale Lake residents. Many of the interviewees intimate that the mourning of botanical loss also involves the mourning of the sensory environments in decline as plants become less common. Posthumanist multispecies mourning, therefore, in part responds to human grief caused by the impoverishing of sensory landscapes for, as Wolfe (2010) argues, posthumanism "opposes the fantasies of disembodiment and autonomy" (xv). For example, Katherine Taylor Smith was born in 1905, lived near Forrestdale Lake until 1917, and passed away in 2004. She recounts childhood experiences from the 1910s and observes the decline of kangaroo paws (*Anigozanthos manglesii*) and other plant species in the area: "They used to grow thick on the sides of the road. It's nothing like that now. Nothing. I mean the fumes from motorcars are what kill everything ... There were many more wildflowers. Kangaroo paws and waxy plants and purple waxes and spider orchids. There was everything. Donkey orchids were still growing in Taylor Road not so many years ago" (in Giblett 2006, 88–9). Conversely, Katherine's sense of mourning implicates the changes caused by exotic species: "In those days there was white sand around the whole of the lake and none of those yangee rushes [*Typha orientalis*]" (89). Moreover, her emotional affinity for Forrestdale Lake is evident in her vivid sensory recollections: "Oh I loved it. I loved the life we had. It's just different, smells different. The country smell was beautiful. The kangaroo paws were the most outstanding smell but they weren't the most beautiful flower" (91).

The qualities of interdependence and connectivity vis-à-vis botanical mourning, evident in these interviews, are further summarized by Don Williams, proprietor of Hi-Vallee Farm in Badgingarra north of Perth. Don describes a form of multispecies mourning, in which the loss of one species has broader implications for the whole system of which it is part, including other plants, animals, and insects: "No one can put a value on individual species. Some people will say if we lose it, it doesn't matter, more will evolve. It would appear that they won't evolve as quickly as we can wipe them out. What a lot of people forget is that if you lose a plant species, you could lose an animal or insect species. And a lot of the orchids have evolved around one individual insect which pollinates them."[4] Don suggests the importance of environmental mourning as a relational response to habitat decline, in distinction to Freudian mourning in which the object in isolation becomes the focus of emotional transference, narcissistic self-identification, and ultimately object-cathexis severance. In sum, the exploration of memories of plants invariably confronts the mourning of the loss of biodiversity. As conservationists, David James and Kim Fletcher have been life-long residents of the Armadale area, and have developed emotional attachments to plants through direct experience of their home places. They express a startling sense of the once-biodiverse Perth suburbs in a process of transformation at the hand of numerous seemingly insuperable economic and environmental forces.

## Conclusion: Transformation through Loss in a Multispecies World

In this chapter, I have outlined three components of a posthumanist model of environmental mourning: relationality, multispecies entanglement, and sensoriality. As the interviews with David James and Kim Fletcher indicate, posthumanist mourning decentres the human without abnegating our emotional, sensory, and mnemonic connectedness. In fact, the posthumanist model developed in this chapter could be said to intensify human expressions of affect toward other beings as human-non-human interstices are valued and made more palpable through this mode of thinking about mourning (i.e. the ecologically situated human subject in their relations). In exploring the interviewees' emotional responses to botanical decline, the model does not fall into the trap of human subjectivity it wishes to transcend. Instead the impoverishment of the interviewees' lifeworlds signifies the loss of their relational situatedness: in the echo of the landscape's emptiness is the voice of the orchid, boronia, or banksia.

Along similar lines, Mick Smith proposes a posthumanist concept of ecological community. Smith (2013, 21) outlines four "materially inseparable" aspects that can assist us in articulating multispecies loss: "material manifestation (appearances), material involvement (effects), semiotic resonance (meanings) and phenomenological experiences." A loss of a species is, therefore, the loss of a species of appearance; a species of creative involvements; a species of significance and openness in the world; and a species of ecological community. In other words, the disappearance of a plant species always entails a rupture of community networks, constituted by the presence of different beings, including plants and humans. Posthumanist mourning attempts to think across the categories of "human" and "plant" altogether, treading a middle way between anthropocentrism on the one hand and ecocentrism on the other.

However, a future-oriented question lingers: How can environmental mourning be made into a resource for constructive change and ecological recovery? Although not in terms of biodiversity loss, Judith Butler has argued that mourning can be leveraged as a "resource for politics" (in Yusoff 2011, 578). Although still an elusive and intangible process by all accounts, mourning, she suggests, "has to do with agreeing to undergo a transformation (perhaps one should say *submitting* to a transformation) the full result of which one cannot know in advance" (Butler 2004, 21). While this transformation can be personal, it can also be multispecies, just as environmental loss reflects non-binary human-non-human entanglement. A posthumanist model of environmental mourning, therefore, opens pathways for recognizing, managing, and recovering from place-based loss so that positive transformation can be catalyzed in response to the severe alteration of home places and habitats. For example, the restoration and conservation of botanical communities returns a vital heritage of plant-based sounds, smells, tastes, sensations, and sights, thereby promoting greater human-non-human well-being in urban areas (Millennium Ecosystem Assessment 2005).

The obverse side of environmental loss examined in this chapter is that activism is in part catalyzed by mourning. The values of posthumanist mourning ensure that transformative actions in response to loss are governed by a concern for plants, animals, and insects and their protection, restoration, or memorialization (Ryan 2012, chapter 11). For example, the Keane Road Strategic Link (KRSL) has been proposed by the Armadale government to bisect Anstey-Keane Damplands, a federally acknowledged threatened ecological community. The road extension would disastrously impact rare and endangered species, such as the critically endangered native short-tongued bee (*Neopasiphae simplicior*) and the endemic graceful sun-moth (*Synemon gratiosa*). As this demonstrates, multispecies thinking is integral to environmental activism campaigns, such as those organized currently by James and the Friends of Forrestdale, and reflects a desire among conservationists to keep intact communities of plants, animals, and insects for the inherent value of those species and the well-being of humans. A posthumanist model of mourning is able to articulate this.

Following Butler, mourning can be an agent of transformation, one that resists anthropocentrism and honours the ecocultural interdependencies between people and place. Returning to where we began, with boronia, the very breath of spring, the continued conservation of this species in its habitat south of Perth should also entail the conservation of its historical and cultural networks, as the means by which human entanglement with plants is nurtured and sustained. While the return of boronia fragrance to the city trains and the call of the streetsellers – "Fresh, sweet-scented boronia! Sixpence a bunch, bor-o-nee-a!" (Parker 1962, 4) – might not be possible now, this should not limit us in imagining new human and plant networks that promote interspecies engagements while also ethically managing plant conservation realities. The human-plant future begins with the simple acknowledgement of these networks – the acknowledgement that plants, particularly of the South-West of Western Australia, are a precious and fragile living heritage. And where there is no possibility of biological restoration, where the effects of habitat loss and disease have run their course, memorialization of lost plant species – in literature, theatre, photography, and painting, and at botanical gardens, environmental centres, government headquarters, homes – can assist communities in mourning their loss.

## NOTES

1 Kim Fletcher, unpublished interview by John Ryan, Perth, Western Australia, 15 April 2013, transcript. All excerpts from interview transcript.

2 David James, unpublished interview by John Ryan, Forrestdale, Western Australia, 23 September 2009. All excerpts from interview transcript, unless otherwise specified.

3 Noel Nannup, unpublished interview by John Ryan, Mount Lawley, Western Australia, 21 July 2010, transcript.

4 Don Williams, unpublished interview by John Ryan, Badgingarra, Western Australia, 28 August 2009, transcript.

## REFERENCES


Albrecht, Glenn. 2005. "'Solastalgia': A New Concept in Health and Identity." PAN (*Philosophy, Activism, Nature*) 3: 44–59.

– 2010. "Solastalgia and the Creation of New Ways of Living." In *Nature and Culture: Rebuilding Lost Connections*, edited by Sarah Pilgrim and Jules Pretty, 217–34. Hoboken: Taylor and Francis.

Albrecht, Glenn, Gina-Maree Sartore, Linda Connor, Nick Higginbotham, Sonia Freeman, Brian Kelly, Helen Stain, Anne Tonna, and Georgia Pollard. 2007. "Solastalgia: The Distress Caused by Environmental Change." *Australasian Psychiatry* 15: S95–8. doi: 10.1080/10398560701701288.

Barad, Karen. 2010. "Quantum Entanglements and Hauntological Relations of Inheritance: Dis/continuities, SpaceTime Enfoldings, and Justice-to-Come." *Derrida Today* 3, no. 2: 240–68.

Beard, John. 1979. "Phytogeographic Regions." In *Western Landscapes*, edited by Joseph Gentilli, 107–21. Nedlands: University of Western Australia Press.

Beresford, Quentin, Hugo Bekle, Harry Phillips, and Jane Mulcock. 2001. *The Salinity Crisis: Landscape, Communities and Politics*. Crawley: University of Western Australia Press.

Bondi, Liz, Joyce Davidson, and Mick Smith. 2005. "Introduction: Geography's 'Emotional Turn.'" In *Emotional Geographies*, edited by Joyce Davidson, Liz Bondi, and Mick Smith, 1–16. Aldershot: Ashgate.

Breeden, Stanley, and Kaisa Breeden. 2010. *Wildflower Country: Discovering Biodiversity in Australia's Southwest*. Fremantle: Fremantle Press.

Butler, Judith. 2004. *Precarious Life: The Powers of Mourning and Violence*. London: Verso.

Clewell, Tammy. 2002. "Mourning Beyond Melancholia: Freud's Psychoanalysis of Loss." *Journal of the American Psychoanalytic Association* 52, no. 1: 43–67. www.apsa.org/portals/1/docs/japa/521/clewell.pdf. Accessed 25 June 2014.

Conservation International. 2008. "Biological Diversity in Southwest Australia." In *Encyclopedia of Earth*, edited by Kevin Caley. Washington: Environmental Information Coalition, National Council for Science and the Environment. www.eoearth.org/article/Biological_diversity_in_Southwest_Australia. Accessed 30 April 2013.

– n.d. "Southwest Australia." Conservation International. www.conservation.org/where/priority_areas/hotspots/asia-pacific/Southwest-Australia/Pages/default.aspx. Accessed 30 April 2013.

Dodds, Joseph. 2011. *Psychoanalysis and Ecology at the Edge of Chaos: Complexity Theory, Deleuze/Guattari and Psychoanalysis for a Climate in Crisis*. Hoboken: Taylor and Francis.

Farr, Douglas. 1999. *Sustainable Urbanism: Urban Design with Nature*. Hoboken: John Wiley & Sons.

Fowler, Jeaneane. 1999. *Humanism: Beliefs and Practices*. Brighton: Sussex Academic Press.

Freud, Sigmund. 1957. "Mourning and Melancholia." In *The Standard Edition of the Complete Psychological Works of Sigmund Freud, Volume XIV (1914–1916)*, edited by James Strachey, 243–58. London: The Hogarth Press and the Institute of Psycho-Analysis.

Freud, Sigmund. 2008. "The Future of an Illusion." In *Philosophy of Religion: An Anthology*, 5th edition, edited by Louis Pojman and Michael Rea, 114–18. Belmont: Thomson Higher Education.

Giblett, Rod. 1996. *Postmodern Wetlands: Culture, History, Ecology*. Edinburgh: Edinburgh University Press.

– 2006. *Forrestdale: People and Place*. Bassendean: Access Press.

Giblett, Rod, and David James. 2009. "Anstey-Keane Botanical Jewel." *Landscape* 24, no. 4: 42–4.

Girardet, Herbert. 2008. *Cities, People, Planet: Urban Development and Climate Change*, 2nd edition. Chichester: John Wiley.

Haraway, Donna. 2008. *When Species Meet*. Minneapolis: University of Minnesota Press.

Homans, Peter. 1989. *The Ability to Mourn: Disillusionment and the Social Origins of Psychoanalysis*. Chicago: University of Chicago Press.

Hopper, Stephen. 1998. "An Australian Perspective on Plant Conservation Biology Practice." In *Conservation Biology for the Coming Decade*, edited by Peggy Fiedler and Peter Kareiva, 255–78. New York: Chapman Hall.

– 2004. "Southwestern Australia, Cinderella of the World's Temperate Floristic Regions." *Curtis's Botanical Magazine* 21, no. 2: 132–80.

Hostetler, Mark. 2012. *The Green Leap: A Primer for Conserving Biodiversity in Subdivision Development.* Berkeley: University of California Press.

James, Jennifer C. 2011. "Ecomelancholia: Slavery, War, and Black Ecological Imaginings." In *Environmental Criticism for the Twenty-First Century*, edited by Stephanie LeMenager, Teresa Shewry, and Ken Hiltner, 163–78. Milton Park: Routledge.

King, Barbara J. 2013. *How Animals Grieve*. Chicago: University of Chicago Press.

Knapp, Sonja. 2010. *Plant Biodiversity in Urbanized Areas*. Wiesbaden: Springer Fachmedien.

Ley, Martin W. 2008. "The Ecological Dimension of Psychoanalysis and the Concept of Inner Sustainability." *Journal of the American Psychoanalytic Association* 56, no. 4: 1279–1307. doi: 10.1177/0003065108327179.

Marder, Michael. 2013. *Plant-Thinking: A Philosophy of Vegetal Life*. New York: Columbia University Press.

McKinney, Michael. 2005. "Urbanization as a Major Cause of Biotic Homogenization." *Biological Conservation* 127, no. 3: 247–60.

Millennium Ecosystem Assessment. 2005. *Ecosystems and Human Well-Being: Scenarios*. Washington: Island Press.

Morton, Tim. 2010. "The Dark Ecology of Elegy." In *The Oxford Handbook of the Elegy*, edited by Karen Weisman, 251–71. Oxford: Oxford University Press.

Paczkowska, Grazyna. 2013. "Boronia megastigma Bartl.: Scented Boronia." *Western Australian Department of Environment and Conservation*. http://florabase.dec.wa.gov.au/browse/profile/4428. Accessed 14 April 2013.

Parker, Pat. 1962. "Boronia Farm." *The Western Australian Women's Weekly*, 24 October. http://nla.gov.au/nla.news-article44027250. Accessed 30 April 2013.

Perth Biodiversity Project. n.d. "Background to the Perth Biodiversity Project." West Perth: Perth Biodiversity Project. http://pbp.walga.asn.au/Portals/1/Templates/docs/background_pbp.pdf. Accessed 30 April 2013.

Pollard, James. 1947. "Boronia." *The West Australian*, 19 July. http://nla.gov.au/nla.news-article46327060. Accessed 30 April 2013.

Porteous, Douglas. 1989. *Planned to Death: The Annihilation of a Place Called Howdendyke*. Manchester: Manchester University Press.

Rose, Deborah Bird. 2008. "Judas Work: Four Modes of Sorrow." *Environmental Philosophy* 5, no. 2: 51–66.

– 2011a. "Flying Foxes: Kin, Keystone, Kontaminant." *Australian Humanities Review* 50: 119–36. www.australianhumanitiesreview.org/archive/Issue-May-2011/home.html. Accessed 30 April 2013.

– 2011b. *Wild Dog Dreaming: Love and Extinction*. Charlottesville: University of Virginia Press.

Rose, Deborah Bird, and Libby Robin. 2004. "The Ecological Humanities in Action: An Invitation." *Australian Humanities Review*, 31–2. www.australianhumanitiesreview.org/archive/Issue-April-2004/rose.html. Accessed 30 April 2013.

Ryan, John. 2012. *Green Sense: The Aesthetics of Plants, Place and Language*. Oxford: TrueHeart Academic Press.
– 2013. *Unbraided Lines: Essays in Environmental Thinking and Writing*. Champaign: Common Ground Publishing.
Shearer, B.L, C.E. Crane, S. Barrett, and A. Cochraine. 2007. "Phytophthora Cinnamomi Invasion, a Major Threatening Process to Conservation of Flora Diversity in the South-west Botanical Province of Western Australia." *Australian Journal of Botany* 55, no. 3: 225–38.
Smith, Mick. 2013. "Ecological Community, the Sense of the World, and Senseless Extinction." *Environmental Humanities* 2: 21–41. http://environmentalhumanities.org/arch/vol2/2.2.pdf. Accessed 1 April 2015.
Smith, Thomas Michael, and Robert Leo Smith. 2012. *Elements of Ecology*, 8th edition. San Francisco: Pearson Benjamin Cummings.
Trewavas, Anthony. 2002. "Mindless Mastery." *Nature* 415: 841.
– 2003. "Aspects of Plant Intelligence." *Annals of Botany* 92, no. 1: 1–20. doi: 10.1093/aob/mcg101.
– 2006. "The Green Plant as an Intelligent Organism." In *Communication in Plants: Neuronal Aspects of Plant Life*, edited by František Baluška, Stefano Mancuso, and Dieter Volkmann, 1–18. Berlin: Springer-Verlag.
Trigger, David, and Jane Mulcock. 2005. "Forests as Spiritually Significant Places: Nature, Culture and 'Belonging' in Australia." *The Australian Journal of Anthropology* 16, no. 3: 306–20.
Tuan, Yi-fu. 1974. *Topophilia: A Study of Environmental Perception, Attitudes, and Values*. Englewood Cliffs: Prentice-Hall.
van Dooren, Thom. 2010. "Pain of Extinction: The Death of a Vulture." *Cultural Studies Review* 16, no. 2: 271–89. http://epress.lib.uts.edu.au/journals/index.php/csrj/article/view/1702. Accessed 30 April 2013.
– 2011. "Vultures and Their People in India: Equity and Entanglement in a Time of Extinctions." *Australian Humanities Review* 50: 45–61. www.australianhumanitiesreview.org/archive/Issue-May-2011/home.html. Accessed 30 April 2013.
*Western Mail*. 1936. "Boronia: Story of Local Perfumery Industry." *Western Mail*, 1 October. http://nla.gov.au/nla.news-article50053018. Accessed 30 April 2013.
Wilson, Edward O. 1986. *Biophilia*. Cambridge, MA: Harvard University Press.
Wolfe, Cary. 2010. *What Is Posthumanism?* Minneapolis: University of Minnesota Press.
Yusoff, Kathryn. 2011. "Aesthetics of Loss: Biodiversity, Banal Violence and Biotic Subjects." *Transactions of the Institute of British Geographers* 37: 578–92.

# 6 Losing My Place

## *Landscapes of Depression*

CATRIONA SANDILANDS

The obscurity of the connections between our own despair and the collective despair that is present in the places where we live adds to our confusion and (political) depression.

ANN CVETKOVICH, 2012

But now she felt, with an almost primordial knowledge, that the first garden must have been a grave.

ANNE MICHAELS, 2009

## Loss, Body, Place

In June 2010 my marriage ended; in November my mother died (both my mother and my marriage had been suffering significant decline for some time). In February 2011 I sold my little house on the Niagara Escarpment; at the end of June, my daughter and I moved into a nice townhouse in the west end of Toronto. The first thing I did was plant a garden, only a fragment of what we had had in the country but a good gesture toward the possibility of taking root in this new place. I transplanted some of my favourite herbs and flowers from my old garden; I also bought rambling roses that would grow in the new, sunnier microclimate of my townhouse. For the first few months, we all seemed to have transplanted quite successfully: new school, new activities and memberships, new opportunities for me to explore and enjoy the city in which I had not lived for seven years. The herbs didn't mind the pots; the roses bloomed prolifically. I looked forward to a winter of not splitting kindling to fire the woodstove in order to stay warm, not worrying about weather conditions for my workday commute, and the general contentment of a new, less tumultuous life.

In November 2011, the infatuation with the newness of our life was over, and the expanse of the future – without my spouse, my mother, my home – became terribly, starkly clear. I fell, descended, sank ("down" is not just a metaphor, it is a physical sensation) – entering into what a clinician later called a "major depressive episode." And much as she also argued that my depression was an understandable response to the multiple losses that I had experienced in the past year and a half – summarized as "adjustment disorder" – my relation to what Samuel Johnson and Winston Churchill chose, more evocatively, to call "the black dog" was anything but reasonable. My body felt thick, my movement slow and viscous, my thoughts unremittingly dull but tinged around the edges with panic whenever I had to make a public appearance, which I mostly avoided; I felt beneath attention. I scraped along, managing to rise to the minimum necessities, but always through a miasma of despair that I simply could not pierce. It took about six months before I could even begin to say I was feeling lighter. I continued to tend to my garden, carefully taking the housewarming-present geraniums in and out to protect them from the winter cold and then expose them to the spring sun, and engaging in more-or-less polite conversation with the dog owners who insisted that my English lavender was impervious to urine.[1] But I was mostly going through the motions.

Trauma and loss are bound up with contemporary understandings of depression. Indeed, despite widespread rejection of most Freudian thought in the world of clinical psychology, his explanation for melancholia continues to carry weight in both clinical and popular discourses. In his view, loss involves the rupture of a libidinal attachment; the beloved object is no longer available for, or even rejects, our love. As a way of preserving that object, we incorporate it into ourselves, internalizing the attachment: the ego holds on to the object by devouring it and making it part of itself, substituting narcissistic for cathectic energy. But that energy includes a great deal of anger: we are furious at the beloved for leaving, and as we internalize the attachment as a way of preserving love, we also internalize the anger (Freud 1984). Hence melancholic depression: we hate the world by hating ourselves. Especially when removed from pathologizing distinctions between normal and excessive mourning (Freud came to consider these internal melancholic attachments as part of a very ordinary process), this story is quite powerful: the person in the midst of depression remains literally possessed by loss, holding tightly to the internalized object, dejected in the midst of her inability to find the border between her own ego and the existence of the loved/hated object within it. Loss is turned inward, and the pain and anger become the ego itself. As Andrew Solomon (2001) puts the whole process: "depression is the flaw in love. To be creatures who love, we must be creatures who despair at what we lose, and depression is the mechanism of that despair. When it comes, it degrades one's self and ultimately eclipses the capacity to give or receive affection. It is the aloneness within us made manifest, and it destroys not only connection to others but also the ability to be peacefully alone with oneself" (15).

Contemporary medical accounts of depression, however, tend to emphasize neurochemistry over existential self-hatred and loneliness. According to Solomon, most research focuses on the role of neurotransmitters in brain function, but there is no clear understanding of causation, and there is an extraordinary lack of accord between what we understand about brain chemistry and the drugs that are most commonly prescribed for treatment. For example, although people in the midst of acute depression do not generally have abnormally low serotonin levels, over time, for some people, taking selective serotonin reuptake inhibitors (SSRIs) that enrich the availability of serotonin in brain for synaptic activity has a measurable positive effect. Hence the occasional dramatic success of Prozac, which also seems to have comparatively few unpleasant side effects to deter therapeutic chemical experimentation (which is what most pharmacological "treatment" of depression actually is, even as it clearly helps some people).

Although the precise understanding of the interactions between monoamine neurotransmitters and other brain chemicals such as neuropeptides and electrolytes remains elusive, the working consensus seems to be that stress interferes with the body's ability to produce one or more neurotransmitters, which are the chemical agents that allow communication between neurons. In this theory, when a person is stressed her neurons fire off very rapidly in order to attempt to solve the stress-inducing problem at hand; in situations where the individual cannot avoid the stress, or where the stress is internally created, neurotransmitters can become depleted over time. The complementarity of this theory with Freud's view is interesting: not only can we imagine that neurotransmitters are a chemical part of the mechanism through which the body incorporates loss, but we can see that the depressed person's particular "lack of peace" has important biochemical manifestations. Grief has a chemistry; prolonged internal conflicts write themselves, in the language of monoamines and glycols, into our neurological composition. We understand very little about this process, however; the rush to "solve" depression with MAOIs (monoamine oxidase inhibitors), SSRIs, and other psychopharmaceuticals has far less to do with a rich understanding of the relations between mind and body than it does with the profit margins of Eli Lilly. More broadly, of course, the problem is also bound up with prevailing reductionist and medicalized models of depression as a disease, and with a neoliberal cultural and political milieu that detests stillness, fallowness, and sadness.

Solomon (2001) notes that the history of the constellation of conditions broadly associated with depression has involved the pendulum of understanding swinging back and forth across the brain/mind divide between medical and spiritual, scientific and philosophical formulations: Hippocratic, Aristotelian, Christian, Burtonian, Romantic, Victorian, psychoanalytic, psychobiological. Indeed, in her fascinating book *The Nature of Melancholy* (2000) Jennifer Radden opens the important question of whether what we now understand as clinical depression is, in fact, the same thing as the disorder identified by the Greeks as melancholia. Black bile included, depression holds a far different meaning in our post-Freudian, neoliberal, medical imaginary than melancholia did in Galen's; if we are being anything other than rigid positivists, we have to ask the question of whether the affective experiences of prolonged despondency, fear, and self-hatred coalesce into the same condition in historical configurations that lie over 1,800 years apart. There is no question, however, that the term "depression" points to a complex landscape that ties together, in historically specific ways, corporeal and environmental, interior and exterior, individual and social processes: it is a nodal point for biopolitical unfolding.

In addition, as Radden demonstrates, up until the late nineteenth century there was not a firm line drawn consistently between the ordinary background state of a melancholic temperament and the particular existence of a disorder of the body, mind, or soul. Nor was melancholy necessarily considered something to be entirely avoided; for example the Romantics believed that a melancholic temperament indicated a state of heightened sensitivity to the world including both pain and pleasure, the sorrowing intensity of the former contributing to the expanded joys of the latter. John Keats's "Ode on Melancholy," for example, includes the observation that "Ay, in the very temple of Delight / Veil'd Melancholy has her Sovran shrine, / Though seen of none save him whose strenuous tongue / Can burst Joy's grape against his palate fine."

Radden's observations are, in this respect, similar to Michel Foucault's in *Madness and Civilization* (1965); where melancholia and other forms of madness were often associated historically with brilliance, states of access to unique forms of knowledge inhabiting the limits of mundane human existence, by the middle of the seventeenth century they were castigated as punishable modes of wilful unreason, and by the end of the eighteenth century they had become dangerous illnesses, requiring protective, putatively therapeutic institutionalization. In a fascinating reconsideration of the condition of acedia, which is associated with early Christian monasticism (and is also much more a matter of spiritual crisis than of melancholic response to loss), Ann Cvetkovich (2012) draws important connections between the restlessness of monks and the politicality of contemporary despair, between loss of faith and loss of hope. The history – as Solomon, Radden, Foucault, and Cvetkovich all show – is far more complex than I can relate here; I do, however, want to raise the question of what it means that people like me experience the constellation of enervation, sadness, anxiety, isolation, despair, heaviness, and hopelessness in the midst of an institutional and discursive context that casts depression specifically as a mental *illness* in a world that fetishizes, even as it does not cultivate, mental *health*.

Cvetkovich (2012) argues convincingly against the prevailing medical model and for a more social and cultural understanding of depression. Asking about the ways in which the organization of emotions is very much part of the biopolitical life of capitalism – what Raymond Williams (1973) calls its "structure of feeling" – she demands that we pay attention to the structural conditions in which depression is so prevalent, such that it is "another manifestation of forms of biopower that produce life and death not only by targeting populations for overt destruction, whether through incarceration, war, or poverty, but also more insidiously by making people feel small, worthless, hopeless" (3).[2] At the same time depression must also be seen "as a category that *manages and medicalizes* the affects associated with keeping up with corporate culture and the market economy, or with being completely neglected by it" (12, emphasis added). The micropolitics of depression, then, also includes making connections between these sometimes debilitating emotional/corporeal experiences and the broader landscape of structures and affects that are part of – indeed, integral to – late capitalism (see also Berlant 2011).[3] To argue for the politicality of depression is not to lessen the import of individuals' experiences of physical and psychical crisis, for some to the point of complete incapacity, hospitalization, and even suicide. I would, in fact, argue the opposite: that the power of these experiences of crisis shows us how very deeply individuals' bodies and emotions, including our complex brain chemistries, are also involved in the unfolding of societal crises.

Of course, one cannot speak of these bodily and social complexities without also including consideration of the *physical* landscape in experiences of depression: places are part of the equation, and losses of place seem especially associated with melancholia. At one level, these losses are about personal change and transition: Cvetkovich (2012) acknowledges that physical landscapes played an important role in her experience of movement, telling her "that a connection to where you are from, especially if it's been denied to you, is crucial" (81). Likewise, in *Where Roots Reach for Water* (1999), Jeffery Smith comes to understand that his depression is strongly linked to homesickness for the Appalachian landscapes that were so much a part of his sense of body and self.

Reflecting on the causes of my own depression, I realized that the loss around which all of the others circulated was the loss of my *place*.[4] I had felt *part* of the rural Niagara Escarpment; I had learned to hear the progress of spring in the changing calls of the crows, the first notes of the returning red-winged blackbirds, the initial "peep" of a single Spring Peeper giving way to a deafening amphibian roar by mid-May (see Bernie Krause's thoughts on soundscapes in chapter 1). I knew the wildflowers, where to find the largest carpets of trout lilies in late April and trilliums in May, where some hiker's boot had allowed delicate helleborine orchids to germinate along the side of a trail. It's not surprising that the first thing I did when I moved in to my new home was to plant a little garden, transplants from home and newcomers together. But even as I was planting, I knew that I was going to be cherishing a very diminished thing: no more of the abundant sorrel and rhubarb that embodied the taste of spring; no more carving the little pumpkins we grew every year for Hallowe'en. The sadness of all of the losses crept into the loss of my garden; the diminishment of my plant community was absolutely visible on a daily basis, and served as a sort of ruin in which my other, less visible ghosts could more easily lurk. In addition, the loss of my garden meant that I no longer had the rituals of digging, planting, weeding, and watering through which to engage in corporeal meditation; I missed my garden not only as a metonym but also as a site of a carefully nurtured physical connection. The little front garden and herb pots made few demands on me, and my body lost its sense of agricultural rhythm. I had lost my place, and part of my self along with it.

But there is more to losing one's place than individual displacement. In her memoir *Even Mountains Vanish*, Sue Ellen Campbell (2003) describes a state in which she "felt trapped somewhere between the blues and outright depression, as if I were encased in some protective kind of suit and helmet that didn't work right, letting in all the world's toxins but cutting me off from everything heartening. Sitting in my yard, I'd think 'What a pretty song that meadowlark has,' then a moment later I'd be brooding about declining songbird numbers and habitat loss" (36). A surfeit of such ecological catastrophies, intertwined with personal ones, for Campbell contributed to a state in which she was not able to see the life in the world, but only the death. Although she eventually arrives at a Buddhist understanding in which precisely the pain of these losses ignites an outward-moving experience of compassion, she also understands what it means to turn the grief inward: "I have read somewhere that the biologist E.O. Wilson has called ours the Age of Loneliness, because we're facing the extinction of so many fellow creatures, and I've thought, *Yes, that's what I'm sensing, that's it exactly*" (94, emphasis in original). Solomon's existential loneliness, then, clearly includes relations to other species; Prozac won't fix the Anthropocene and, with Deborah

Bird Rose (2013), we must "*howl* in the dark for the loss that surrounds us now, and for all that is coming" (1, emphasis added).

I have written elsewhere (Mortimer-Sandilands 2010) about the ways in which this kind of melancholic understanding may be an important tool with which to respond to environmental losses. In particular, following in the footsteps of AIDS activism with its emphasis on the politicization of mourning, one can imagine environmental melancholia as a condition that promotes an acknowledgement, as Campbell does, of the personal and affective dimensions of what is being lost. We have few public rituals for the loss of places, species, or ecosystems; in the face of everything from suburban sprawl to the Alberta Tar Sands, Judith Butler might call such losses "ungrievable" (2004). And so, perhaps, we turn them inward; in a last effort to hold on to a Labrador duck, a North African elephant, a Xerces blue butterfly, or a woolly-stalked begonia, we make their loss part of ourselves, its trace in our sadness the last sting of the English short-haired bumblebee. And rather than simply turn our attentions and attachments to the new, as consumer capitalism would have us do (and in as unimpeded a manner as possible), we bring our griefs with us as we think toward the future. A politicized depression is, here, one that insists on recognizing and indeed acting from despair. By allowing our lost places to be fields of *grieved* lives and possibilities, we might enable more thoughtful reflection on what it is we are losing, and what we lose about ourselves in the process.

## Writing, Madness, Landscape

In her brilliant book *Writing and Madness* (1985), Shoshana Felman, negotiating between Foucault and Derrida, considers the relations among madness, philosophy, and literature. As she describes, for Foucault the cultural history of madness has been one of its increasing exclusion from reason. The enunciative and denunciative task of the historian of madness is, then, "that of finding a language: a language other than that of reason, which masters and represses madness, and other than that of science, which transforms it into an object about which no dialogue can be engaged, *about* which monologues are vacantly expounded – without ever disclosing the experience and the voice of madness in itself and for itself" (41). For Derrida, in contrast, "the relationship of mutual exclusion between language and madness ... is not *historical*, but *economical*, essential to the economy of language as such: the very status of language is that of a break with madness, of a protective strategy, of a difference by which madness is deferred, put off. With respect to 'madness itself,' language is always *somewhere else*" (44). And as Felman formulates it, this "silence of madness ... is not said in the logos of the book but rendered present by its pathos, in a metaphorical manner, in the same way that madness, inside of thought, can only be evoked through fiction" (47).

Literature is thus the meeting place between madness and philosophy. But the very performative, evocative qualities of a work of literature that allow the reader an experience of the "somewhere else" of madness are themselves contingent on a form of madness, namely, the hallucinogenic properties of the literary work itself. Fictions of madness simultaneously deny it and offer it; where on the one hand to talk about madness is always to displace madness itself, on the other hand, there still exists in such talk "a *madness that speaks*, a madness that is acted out in language, but whose role no speaking subject can assume" (252, emphasis in original). This is a madness of literary rhetoric itself. Or, as Felman puts it finally: "literature's particular way of speaking about madness consists in its unsettling the boundary ... between 'the madness that gets locked up' and 'the hallucination of words'" (253). Thus, fiction not only challenges the limits of institutional discourses of madness, but also crafts, for both writer and reader, an experience of the porosity of sanity itself, a fleeting sense of the daily mechanisms by which both the power relations and the economies of language puff up the speaking subject against the swirling demons that haunt the borders of representation.

There is no shortage of literary works about depression, understood here as a form of madness; indeed, the romantic association of melancholy with genius would appear to be borne out by the number of writers attempting to craft their own madness into literature (Keats, Virginia Woolf, Sylvia Plath). In addition, however, a large number of so-called depression *memoirs* have emerged in recent years by authors ranging from William Styron to Sally Brompton to Les Murray. Despite their differences, these works form a distinct genre: they are intimate accounts of at least one major period of depression in the author's life.

Most also include several other elements in common: detailed retrospective description of an individual's process of "breaking down"; an attempt to understand and narrate the cause of their depression, usually focusing on a series of major losses but often also including *ex post facto* revelation of a series of prior minor depressions; often, the individual's own more-or-less scholarly research into depression (Solomon's *The Noonday Demon* is the most thorough example); a description of a slow "recovery" process, almost inevitably offering advice to others on the relative or combined merits (or not) of talk and drug and even electroconvulsive therapies; and, in the most thoughtful works, a lingering sense of what it feels like to look *ahead* to a life with depression, rather than just backward to its defeat.

As Radden notes (2009), some depression memoirs are what she calls "symptom alienating" – the author composes and positions her sense of self as outside the condition of depression, and recovery is depicted as a process of regaining that self from the grip of an external force. And as Cvetkovich might add, these are also the accounts that most clearly treat depression as a pathological, individual condition to be overcome by any means possible. Other memoirs, perhaps also connecting to less medicalizing traditions of thinking about despair and melancholia, are what Radden calls "symptom integrating," in which the author comes to understand depression as an important part of the self, and in which the experience of depression is understood as transformative, meaningful, and valuable.

In a striking example of the latter, in *In the Jaws of the Black Dogs* (1995) John Bentley Mays comes to recognize his own depression as a corporeal embodiment of the world's malaise, a painful and living side effect of consumer capitalism and hyperindividualism. Although Mays certainly considers his depression as an affliction, his experience is entwined with a more pervasive *societal* affliction, and his breakdown in the midst of these conditions has sharpened his insight into a larger malaise. He writes, for example, that his depression did not: "return me to a state of primordial innocence, nor transform me into a happy consumer-citizen of mass society. It *did* smash bad communications inside me and with others, and reshuffle the deck of images that constitute the *I*. It reordered my sickness so that, most of the time, I can get on with an ill brain, and in a world in dis-ease. Perhaps I am not 'cured' ... because I do not want to be; to live on in that ignorance of the suffering world would be to live as a zombie .... If I would like to be cured of anything, it is the individualism inscribed in me by mass culture, and present always as moralizing, medicalizing scrawls on the walls of my brain" (225–6).

Reflecting on the process of writing the memoir itself, he understands the work as a way of helping to "rewrite" depression: "construct[ing] new scripts so that the old, shabby ones could be discarded, or at least edited to make them easier to perform without a stammer" (184). Mays is, in a way, writing his madness into the world not as an act of dispelling the stigma of depression – as many memoirs claim to be doing – but rather as an act of rendering his madness a symptom of the world, representing it faithfully so that his own depression can be read as an acute instance of a larger problem. His writing of madness is thus a thoughtful opening to the world even as his experience of depression was a horrific closing: "I experience both depression and its treatment as continuous mass-culture phenomena, not individual disease and one-on-one therapy. Depression has always been for me, and remains, a self-punishing language, a prolonged sensation of filthiness and worthlessness, of embarrassment at being alive; a sickening deadness I enviously compare to the liveliness other people seem to enjoy. I cannot imagine it other than as a constellation of images and words inextricably connected to models, disturbances, languages "out there," in the field of social existence" (216). With Felman in mind, I suggest that some depression memoirs (such as Mays's) can thus be read as a form of public sense-making of an experience that, in the moment, cannot possibly make sense publicly. Such is the nature of acute depression that it is irretrievably singular, that at least part of the self-hatred in experiences of depression is centred on the gulf between the excruciating now of the incapacitated *I* and the ongoing plenitude of everyone and everything else.

Although some memoirs – such as Cvetkovich's thoughtful and complex rejoinder to Prozac-fuelled "I conquered it" stories – contain the wisdom that depression is a matter of process rather than event, of grades of life rather than a state of illness, there is also in many of them a point at which the actual experience of a depressive breakdown sits perpetually at the edge of intelligibility. It cannot be written at the moment of its experience, but can be retrieved only as a past condition from a present and relatively communicative lucidity. The conundrum is, then, that the words are always already a concession to a hallucination of representability, and it is only with a huge faith in the pedagogical or other power of that hallucination that the depressive can possibly write of her or his own experience at all.

With Mays also in mind, what interests me is the sense that the evocation of depression may form a mode of social critique: the attempt to place into language the experience of acute depression, here, constitutes a move to perform the corporeal and psychic pathologies of the world. The narration of a so-called major depressive episode does not tell readers what to think (Mays does so later), but instead asks us to consider feeling the world in a particular way, to connect the experience of the text to the experience of the world, *both of which are mad*. Most interesting, then, are the ways in which depression memoirs characterize the experience of the so-called "depths" of depression, offering up the irreducible singularity of the worst moments to the (im)possibility of shared understanding. Here is one example, from the memoir/poetry collection of Les Murray, *Killing the Black Dog* (2009): "Every day, though, sometimes more than once a day, sometimes all day, a coppery taste in my mouth, which I termed intense insipidity, heralded a session of helpless, bottomless misery in which I would lie curled in a foetal position on the sofa with tears leaking from my eyes, my brain boiling with a confusion of stuff not worth calling thought or imagery; it was more like shredded mental kelp marinaded in pure pain. During and after such attacks, I would be prostrate with inertia, as if all of my energy had gone into a black hole" (7). The phrase "like shredded mental kelp marinaded in pure pain" has no particular instructive intent; it is simply – and yet not at all simply – an attempt to bring the reader into a precise experience of anguish. Murray acknowledges that his experience exceeds his description: his thoughts are "like" shredded kelp, and the taste in his mouth is "termed" by him as intense insipidity. He is also watching himself – from the writerly present – as he curls into a foetal position, tears leaking.

Murray thus exposes his own ongoing struggle to give a voice to madness; the *I* that is writing is not the same as the helpless *I* on the sofa. Nevertheless, he evokes a particular taste, an imagistic cacophony of mental garbage, and an episodic rhythm of coppery beginning, excruciating middle, and exhausted end that are only and always part of the interior of the depressive experience; evocation, here, emerges in the gaps left open in simile and reflective attitude. And importantly, the evocations are directly sensuous: a taste, a mental image, a sense of movement. The attempt to render depression palpably present is, then, partly metaphoric – as it cannot help but be – but it is also partly a matter of experience, an opening of the reader (and the writer) to an inarticulate depressive interiority directly through the senses.

Among other things, then, depression-writing – memoir, but also poetry and fiction – can add this kind of sensory immediacy and reflexivity to our understanding of environmental loss and anguish. In particular, I would like to suggest that the evocation of places and natural processes *as melancholy* in literature can be read not only as a matter of the projection of depression onto a landscape by a melancholic writer, but also as a sensitivity to the despair that is part of the environment itself. Radden (2009) asks about the ways in which melancholia is attributed to non-human forms and beings such as a bleak, grey moor or the lonely hoot of an owl. On the one hand, of course, these attributions are part of a deeply historical and widely disseminated series of visual and aural stereotypes: we have learned, through repetition, to make associations between the natural world and the moods and conventions through which we apprehend that world, in this case, darkness and isolation. On the other hand, however, even if we are always conditioned by such cultural textures, "we [also] attribute expressive properties such as melancholy to natural phenomena because those phenomena directly induce particular feeling states in us" (186); there are, in other words, particular resonant qualities to certain landscapes that are more likely to induce in us anxiety, sadness, and loneliness, perhaps because we can see the losses and traumas that constitute us all, together.

Take, for example, the following excerpt from Sylvia Plath's poem "Wuthering Heights":[5]

> The horizons ring me like faggots
> Tilted and disparate, and always unstable.
> Touched by a match, they might warm me,
> And their fine lines singe
> The air to orange
> Before the distances they pin evaporate,

Weighting the pale sky with a soldier color.
But they only dissolve and dissolve
Like a series of promises, as I step forward.

Here and throughout the poem, we are faced with Plath's powerful sense that the landscape is an unpleasant antagonist: something disliked and bleakly ugly, but also something materially discomforting and disorienting. The hills that surround her are ineffable – "tilted," "disparate," and "unstable" – but with none of the romantic overtones of awe or wonder; as she goes on to write, "there is no life higher than the grasstops / Or the hearts of sheep." Neither is there comfort in the warmth she imagines the hills might provide: the image of Plath ringed by faggot-like horizons speaks more of witchburnings than it does the glow of a jolly communal blaze or the familiarity of a signal fire on a distant ridge.

Still, even that violent warmth eludes her: the horizons "dissolve and dissolve" as broken promises along her journey, evoking and yet exceeding the everyday experience of being lost: her disorientation is not ordinary, not reparable, and each stanza extends our sense of her discomfort. The sheep, actually menacing inside their "grandmotherly disguise," peer at her through "the black slots of their pupils ... / It is like being mailed into space, / A thin, silly message." And in the final stanza, "the sky leans on me, me, the one upright / Among all horizontals"; she is the only thing left standing amidst all of this wuthering bleakness, and even the air has forgotten what would, for Plath, be the sound of human language, except, perhaps, for the voice of the grave: "black stone, black stone." And almost in the centre of the poem, there is the kernel of death around which all of this alienation circulates: "If I pay the roots of the heather / Too close attention, they will invite me / To whiten my bones among them." The only thing attractive, in the midst of this utter bleakness, is the thought of interment.

This landscape of Plath's goes beyond a mere brooding sadness; it has taken on the character of all the dissolved promises of her life. It is a tortured place in which only the ugly undersides of things are exposed, in which fluffy sheep are grey clouds, wind-tossed grass is beating its head in distracted self-punishment, and even the promise of fire has no actual warmth. It is a fierce anti-pastoral:[6] all joy is sucked out of what to other minds might be a beautiful scene. It is *depressed*: the grass is frightened of the darkness, and the siren suicide-call of the heather roots appears as a welcome option in these narrow black valleys where the house lights are simply "small change."

Plath has brilliantly evoked the experience of depression as if it were a quality of the landscape itself: in much the same way as my own losses were catalyzed by my unhappiness in my new, diminished place, Plath gives her despair *to* the landscape as if the affect originated there (in part, perhaps it did). Some critics have accused this poem of extreme egoism; I would argue that the simultaneity of ego and place is actually part of its point. It is an interior landscape rendered faithfully as an exterior one. And so its anti-pastoralism is critical: where readers might expect fullness, plenitude, mauve hills, and gentle sheep, we are only offered the very darkest possibility of each. We know from (neo)pastoral expectation that this place should be full of vitality, but we also know that that experience is rendered impossible in the poem. In other words, the poem evokes for the reader an experience of extreme depression by using literary landscape conventions against themselves: we *feel* the denial of plenitude in the space between Plath and the pastoral; we *sense* the loneliness of the grave-reciting wind; we *suffer* the disenchantment of the dissolving horizons.

Like Mays's assessment that his depression is part of the pathology of the world, Plath's frankly depressed landscape enacts not only the underside of the pastoral, but also the rottenness of a world for which the pastoral is an alibi.[7] Specifically, I think she gives us a *precise* sense of what it means to be alienated from the more-than-human world; if the "normal" relations to the world that are required of everyday life in neoliberal capitalism require also that we play along with the pastoral fantasy, then (intentionally or not) Plath slices through the fantasy to show something much more painful: *that the world is in despair*. Of course, the poem is not a direct critique of neoliberal capitalist nature relations; what it does, however, is evoke such an intense experience of alienation from/in body and place together that the reader cannot help but pause to feel that there *is* an unexpected darkness there. Although the precise subject of the poem may be very particular, perhaps its emotional resonances are not: perhaps her evocation of an actively despairing world can touch off something in our reading experience that will open us to the madness that is at the centre of the capitalist landscape.

## Loss, Place, Writing

Anne Michaels's novel *The Winter Vault* (2009) begins on the site of the temple of Abu Simbel, during the process of its being painstakingly taken apart and moved uphill in advance of the 1964 damming of the Nile to create the reservoir Lake Nassar. Jean, a Canadian, is there with her husband Avery, the British engineer in charge of the massive and exacting task of cutting and parceling the temple, piece by piece, and reconstructing it exactly in a location safe from the imminent deluge. The Nubian people who cohabit the Nile with the temple are also being systematically taken apart and relocated to sterile, hastily-built towns hundreds of miles away: in 1964, "Nubia in its entirety – one hundred and twenty thousand villages, their homes, land, and meticulously tended ancient groves ... vanished" (16). Indeed, as Michaels adds, "even a river can drown; vanished too, under the waters of Lake Nassar, was the Nubians' river, their Nile, which had flowed through every ritual of their daily life, had guided their philosophical thought, and had blessed the birth of every Nubian child for more than five thousand years" (ibid.).

Jean and Avery had met on another of his massive engineering projects, the building of the St Lawrence Seaway in 1954. Jean was collecting plants from the riverbank to preserve "these particular plants, this particular generation" from a place about to be flooded, "to keep a record ... though of course they'll never grow and reproduce themselves exactly as they would have, if they'd been left alone" (50). As they fall in love and tell each other the stories of their lives in a cabin at Long Sault, Avery shares with Jean his love of engineering, bound up with love for his father, with whom he had both travelled as a child on projects and later apprenticed. In return, she shares with him her love of plants and gardens. In particular, she tells him about the loss of her mother when she was a child, how she had taken cuttings and plants from her mother's garden, and how these plants embody both her mother and her life task of dealing with her loss: "my botany, my love and interest in everything that grows – at first it was for love of my mother, a way of living with my yearning, and then perhaps an homage, but gradually it became something more, a passion, and I wanted to know everything: who had made the first gardens, how plants had been depicted in history, growing up in the cracks of cultures, in paintings and symbols, how seeds had travelled – crossing oceans in the cuffs of trousers" (61–2). For Jean, then, places, and especially gardens, are melancholic embodiments of memory: plants carry relationships both symbolically and in their genetic codes, gardens are places in which to preserve or recreate relationships,[8] and botany and gardening represent a passionate commitment to tending these relationships (she later replants the cuttings from her mother's garden in that of her mother-in-law, Marina). Hence, as she is preserving specific common plants from the banks of the St Lawrence, she is also attempting to preserve an entire culture about to be moved or drowned.

As Michaels records, both megaprojects were ecological as well as social disasters: the waters of Lake Nassar evaporated at a fantastic rate; the silts that had enriched the fecund Nile floodplain were cut off, which also affected the Mediterranean fishery; there were massive downstream erosions, species extinctions, and infestations, and a corresponding rise in insect-borne diseases. An entire people was dispossessed, gutted of their connection to place. The St Lawrence Seaway displaced many white settler towns and villages; houses were moved, but the places where they had grown could not be, and the lives that had been sustained on farms and in towns were also detached and destroyed. Graves and gardens alike were drowned. In addition, "the First Nations ... were dispossessed of shore and islands, and heavy metals from the new seaway industries would poison their fish supply and their cattle on Cornwall Island. Spawning grounds [were] destroyed. Salmon would struggle upstream, alive with purpose, to find their way blocked by concrete" (39).

For Michaels, then, losses are carried in places as part of their (and our) embodied memory, and losses *of* places – in the novel, entire landscapes drowned or otherwise obliterated – are also written on the body in especially traumatic ways: habits and modes of inhabitation are unraveled, senses are disquieted, ecological and kinaesthetic connections are severed, relationships and rituals wither and die for lack of the right soil and light.[9] Melancholia is not, here, a simple question of personal pathology; it is a complex of relations in which people are bound up in processes of destruction, violence, dispossession, and dramatic ecological decline that register their effects on both individuals and landscapes. Loss is *in* landscape; landscapes *do* despair. For example, when young Avery is with his father on a trip to Turin, he glimpses a wooden board at the train station noting that the deportations of Jews had taken place from that place "and gave the number – in the hundreds of thousands – of those who had been sent to death from the very place we stood" (105). Avery's entire experience of the city is permeated by the lingering, horrific sorrow of this history. And as he later asks Jean, "Would the city have felt ominous to me even if I had not seen that marker in the station? Would I have felt the foreboding nonetheless, this presence, this dread, this hauntedness we sometimes feel – inexplicable, ineffable – in certain places, in a cast of light?" (106). Clearly, both he and Jean think so.

While Jean and Avery are in Nubia, Jean becomes pregnant with a daughter, Elisabeth.[10] In late pregnancy, the baby dies (Jean has just dreamed of another child's death); according to medical wisdom of the day, she carries Elisabeth to term, in the interim reflecting on the relationship between her daughter's corpse and her own body: a grave. As Jean sinks into incapacitating depression, Avery completes the salvage of Abu Simbel in a Nubia now devoid of its people: he "was haunted, the desert was haunted, by the emptiness of the villages, by their destruction, by impotence and mourning, by the lie of the replication" (176). As Jean's body mourns the loss of Elisabeth, "moon and tide" (173), so too does the desert despair for its children, as "Avery felt Jean's suffering, in the ache of the cliff, in the silent villages" (176). But even as Jean returns to her mother's flourishing garden in Ontario, she cannot find solace in the soil; she carries a new sadness in her body – "as if she were used to stooping from fatigue, from futility" (188) – that cannot be lifted, even by the rituals of growing and tending plants: "she worked until her hands ached, until the low, intense glow of twilight spread across the garden. She longed for the first clarity of autumn, wondering if the cold could cleanse her. But she knows it could not" (189–90). Jean, steeped in depression, *cannot* find meaning and pleasure in the fecund futurity of plants or gardens – "she did not understand what her botany meant to her now, nor what to do with it" (193) – and finds herself absorbed, instead, with tracing the materialities of *loss* as they transect and connect body and place.

Although there is much more to the novel, one thread that Michaels carefully follows through intertwined personal and ecological catastrophe concerns the porousness of the relationship between person and place in the experience of depression. Even more than Plath's menacing hills, perhaps, Michaels's aching cliffs are not simply a projection of human emotion onto a place but instead show the anguished qualities of specific landscape histories; at the same time, though, for Michaels depression crafts in individuals ways of relating to places that unearth death rather than fertility, that allow a *seeing* of the traumas and dispossessions that have brought us and our places into being. *The Winter Vault* is quite extraordinary in its dense interweaving of trajectories of despair – the Holocaust, engineered environmental catastrophes, deaths, displacements, and shattered hopes craft unlikely communities of broken people who connect across losses – and it is a difficult, sometimes exquisitely painful experience to read the novel as a result. But through Jean's despair we *feel* the destruction of Nubia; through the eyes of the villager who refuses to have her husband's grave relocated, we *feel* the intimate as well as the large-scale violences of the damming and draining and redirecting of the St Lawrence; and through the rememberings of hills, rivers, cities, and deserts, we *feel* the ways in which our corporeal histories are woven into multiple violences of place. Unlike most memoirs, there is not an "after" in this novel, a moment from which one can reflect on a depressed past; there is only an ongoing, unfolding dialectic of complex embodiments of loss and devastation, to which we can respond more or less well either personally or societally.

Of course, the other thread that Michaels traces specifically concerns gardening as an embodiment of these relations: as Jean transplants cuttings from her mother's garden into Marina's, she roots the loss of her mother at the centre of this new relationship; as the Nubians and southeastern Ontarians lose their gardens to engineered floods, they lose parts of their daily embodied identities; as Jean is alienated from her despairing body, she is unable to cultivate solace by working her hands into the earth. But when Jean moves back to Toronto, a few blocks from Avery during their separation, she begins to understand gardening in a new way. Specifically, she begins to plant memories for Toronto's immigrants; surreptitiously, at night, in unclaimed public spaces, she plants "cuttings that would grow unnoticed except for their fragrance ... [so that] familiar scents would invade their dreams and give them inexplicable ease" (196). But this act of cultivating embodied landscape memory does not bring her fulfilment (not even the physical "meditation of lifting the earth one scoopful at a time" [197]); "after a night of planting, she was stunned with loneliness, as if she's been tending graves" (196). For of course she has been: for Jean and many others in the novel, flower gardens not only carry the memories of the lost (as with her mother's garden), but they are also planted or left specifically to commemorate the dead, even to *invoke* them. Thus Jean's flowers smell of home, and also of death and displacement: she is planting graves for lost places. And perhaps in recognition, on the first anniversary of Elisabeth's death, she takes seeds she collected from the banks of the now-flooded St Lawrence, on the day she met Avery, and plants them on her daughter's grave, all the losses coming together in the grave, the grave that was the first garden.

When I planted my new little garden in Toronto, it was with hope, thoughts of the future and fertility. Although I understood as I did so that it was a space of some diminishment, I did not comprehend until many months later that the garden was literally a memorial to my dead; not surprisingly, re-reading *The Winter Vault* in spring 2011 helped catalyze this understanding. It was not that my loss of place summarized all of the other losses of my recent past; it was that it *summoned* them, focused them, corporeally and emotionally reminding me of the relationships, smells, and practices that were no longer there (in the garden, the lavender particularly evoked my mother). Crucially, it was not until I let go of the fantasy of the garden as a gesture toward plenitude that I began to understand the deaths and losses as they had registered in my body; accepting my own constitution by these losses in turn allowed me to eventually begin to make new connections to others' despair, both individual and collective.

I do not particularly wish to say that depression is a sort of tragic sensitivity to the ills of the world, ecological and otherwise. But it remains the case that depression offers a sort of limit-experience: not just an excess to reason, qua Felman, but also an excess to the affective relations of capitalist modernity, including relations to the places that are our primary, everyday experiences of the natural world. Depression includes a strong sensation of banishment from the everyday plenitude of the world, from precisely the optimism of vitality, growth, and renewal. It is an alienation from eros, a feeling of separation from life, and thus a place from which to claim to know the underside of those values: enervation, disintegration, death. Environmentally, we need to pay attention to this underside; Michaels and other critical writers of depression do us a great service by allowing us to feel the anguish that such attention affords, thereby offering us an important tool with which to intervene in the emotional ecologies of late capitalism.

## Epilogue

In her essay "In the Shadow of All This Death" (2013), Rose argues that the Anthropocene demands of us a sort of "crazy love," an ethical and emotional abandonment to the possibility of intimate attachments with creatures – in her case, Australian flying foxes – who are almost certainly doomed to perish as a result of climate change, habitat loss, chemical contamination, and/or any of the host of other anthropogenic processes that challenge so many other species' ability to survive. "It takes crazy love," she writes, "to keep defending the lives of the persecuted, and over time it puts us in a place of witness to the apparently unstoppable and the increasingly unimaginable. This is a place of emotional turmoil and exhaustion. In the eloquently understated words of one flying fox carer: 'we are not holding up very well here'" (17).

Crazy love emerges from our capacity to respond generously to the Other. For Rose, following Emmanuel Levinas, "one only comes into becoming within the entangled worlds of life and death through others and through one's response and responsibility to others" (11); generous response means attentiveness to the losses that inhere in becoming, ours and others'. Coming into becoming inevitably means coming into the realm of death as well as life; what Butler (2004) terms our "shared precarity" means that the condition of all of our earthly lives is essentially one of death and, with some ethical and political work, of the survival of love and community in its midst. Depression is a particularly *crazy* part of crazy love, of a passionate corporeal recognition of our collective precarity; it is an unbearable condition of a body and mind that has extended itself to the Other to such an extent that the loss of the Other entails a dramatic change to the self, a madness of becoming.

Although again it is important not to romanticize it, it is nonetheless worth considering that depression is an important affect for the Anthropocene: not a despair to be quietly medicated or talked away, but rather a significant and powerful experience of our interconnectedness with others, perhaps involving forms of love that we are not able to articulate fully until after they are gone, as with the irretrievable loss of a place or species or people. Loss connects us in ways that plenitude does not; in this respect, depression in the Anthropocene demonstrates a vulnerability to loss that also lies at the heart of renewed multispecies communities, ethical commitments, and forms of attachment that are not satisfied by diminished replacements. In order to mourn we must love; in order to love, I think, we must mourn.

I did not know exactly how important my old garden was to me until I tried, and failed, to resurrect it in miniature: until I recognized that the columbines and lilies and echinacea and asters that I had brought with me "from home" spoke of absence as much as presence; until I acknowledged that I *missed* the time-consuming weeding and watering (a ritual tending that turned out to have been a very crazy love indeed, as the people who bought my house almost immediately put sod over the vegetables and herbs that I had left for them). But my eventual (if unwilling) embrace of these absences allowed me to experience the city differently as I began to recognize not the apparent plenitude of others but rather the ways we are connected by our losses and, only then, by love. Most pointedly, I began to volunteer for a local park – quite an ecologically degraded space in comparison to where I had lived on the Escarpment – among other things leading guided group walks to help raise awareness of the animal and plant life that *persists* in Toronto in the midst of pollution and habitat loss. I lost my place, yes: but I learned to love the snapping turtles. While I still can.

## NOTES

1 See my short essay "Lavender's Green? Redux."

2 To say that depression is "prevalent" implies that everyone is equally prone to feeling or reporting it. Such is not the case; indeed, it is a profoundly biopolitical affect, in which there is a complex relationship between depression and social inequality. Women are more likely to experience depression than men (although some groups of women are much less likely to report it); poor people are more likely to report depression than the wealthy (but they might also have to identify with depression as a clinical disease in order to qualify for health coverage); people living alone have

a much higher incidence of depression than others (which might indicate that isolation contributes to depression, but also might suggest that having a resident network of friends and family lessens the need to turn to medical professionals); and, in Canada, Indigenous peoples (both on and off reserve) have a very much higher rate of depression than anyone else, underscoring the fact that hopelessness, isolation, and despair are not at all equally distributed, that trauma and loss are often collective (including intergenerational) experiences, and that treating depression as a question of individual pathology misses the point for many, many people.

3 Cvetkovich also draws important connections between depression and racism, much as earlier feminists pointed to its connections to gender.

4 My original home was in Victoria, BC, which I left – like both Cvetkovich and Smith – for academic pursuits in 1987. I had never intended to leave permanently, but I ended up getting a job in Toronto and although I tolerated the city, I had felt in a state of semi-permanent exile from the possibility of being "at home" until I moved to Caledon. The home I experienced there was not idyllic, but nonetheless I developed a deep sense of connection to the specific plants, animals, and even rocks that were my daily companions. Imagining place and loss together need not be a matter of nativity and origin, although it is *very* important, as Cvetkovich indicates, in settler-colonial landscapes in particular, to connect both personal and environmental losses of place to those of *displacement and dispossession*.

5 Plath's 1963 (1999) *The Bell Jar* also includes the following much-quoted attempt to describe depression directly: "I felt very still and very empty, the way the eye of a tornado must feel, moving dully along in the surrounding hullabaloo" (3).

6 Of course, it is also gothic, especially in its obvious invocation of Brontë. Several recent critics have explored the importance of gothic traditions to environmental literature, including Lousley (forthcoming); I would consider that there are clear affective resonances between those projects and mine here.

7 On the neopastoral as a distortion of capitalist rural/urban relations in particular, see Williams's unrivalled *The Country and the City* (1973). My riff on his central argument would be that the suburbanized fantasy of the neopastoral actively works to diffuse the experience of environmental loss. Naming a tidy cul-de-sac in Brampton or Langford after the landform that has been destroyed (or the species extirpated or the people displaced) to make way for lawns and houses is a bad-faith way of preserving the image of the dead without ever having to feel its loss, and thus of affirming the bourgeois naturalness of the present without ever having to consider the relations of destruction that went into its creation.

8 Michaels also dwells on the ways in which Nubian gardens in the Nile floodplain involved intricate networks of responsibility and entitlement toward different kinds of date palms (15, 20–1), part of the rich tangle of place, embodiment, and sociality that was part of the Nubian people and that was completely destroyed when the people were dispossessed.
9 The first part of the novel centres on Jean's relationship with Avery, up to the death of their child, their return to Toronto, and Avery's attempt to shift Jean's despair by allowing her her "freedom" in a separation (he to study architecture, she to pursue botany). The second part centres on Jean's relationship with Polish Jewish immigrant artist Lucjan, who escaped the ghetto and experienced the destruction and rebuilding of Warsaw as a child. Lucjan tells her his story (as had Avery); his is, like hers, centrally concerned with the relations between personal losses and losses/reinhabitations of place. For brevity, my discussion in this paper primarily concerns Jean's relations to loss. But Lucjan also comments on flowers, noting that the first shop to open up in the ruins of Warsaw was a florist: "no one said what was surely simply and obvious: you need flowers for a grave. You need flowers for a place of violent death. Flowers were the very first thing we needed" (217–18).
10 Elisabeth is also the name of Jean's mother; Jean is not able to redeem her mother's death through the birth of her daughter.

## REFERENCES


Berlant, Lauren. 2011. *Cruel Optimism*. Durham: Duke University Press.

Brompton, Sally. 2008. *Shoot the Damn Dog: A Memoir of Depression*. New York: W.W. Norton & Co.

Butler, Judith. 2004. *Precarious Life: The Powers of Mourning and Violence*. London: Verso.

Campbell, Sue Ellen. 2003. *Even Mountains Vanish: Searching for Solace in an Age of Extinction*. Salt Lake City: University of Utah Press.

Cvetkovich, Ann. 2012. *Depression: A Public Feeling*. Durham: Duke University Press.

Felman, Shoshana. 1985. *Writing and Madness: Literature, Philosophy, Psychoanalysis*. Translated by Martha Noel Evans. Ithaca: Cornell University Press.

Foucault, Michel. 1973. *Madness and Civilization: A History of Insanity in the Age of Reason*. New York: Vintage Books.

Freud, Sigmund. 1984. *On Metapsychology: The Theory of Psychoanalysis*. Translated and edited by James Strachey and Angela Richards. Penguin Freud Library, vol. 11. London: Penguin Books.

Keats, John. 2014. "Ode on Melancholy." Representative Poetry Online: University of Toronto Libraries. http://rpo.library.utoronto.ca/poems/ode-melancholy. Accessed 14 October 2015.

Lousley, Cheryl. Forthcoming. "Diseased Nations, Gothic Ecologies: *Silent Spring*, *Surfacing* and *Salt Fish Girl*." In *Green Words/Green Worlds: Environmental Literatures and Politics*, edited by Catriona Sandilands, Ella Soper, and Amanda Di Battista. Waterloo: Wilfrid Laurier University Press.

Mays, John Bentley. 1995. *In the Jaws of the Black Dogs: A Memoir of Depression*. Toronto: Penguin Books.

Michaels, Anne. 2009. *The Winter Vault*. Toronto: McClelland and Stewart.

Mortimer-Sandilands, Catriona. 2010. "Melancholy Natures, Queer Ecologies." In *Queer Ecologies: Sex, Nature, Politics and Desire*, edited by Catriona Mortimer-Sandilands and Bruce Erickson, 331–58. Bloomington: Indiana University Press.

Murray, Les. 2009. *Killing the Black Dog: A Memoir of Depression*. New York: Farrar, Straus and Giroux.

Plath, Sylvia. 2008. *The Collected Poems*. Edited by Ted Hughes. New York: Harper Perennial.

– 1999. *The Bell Jar*. New York: Faber and Faber.

Radden, Jennifer. 2000. *The Nature of Melancholy: From Aristotle to Kristeva*. New York: Oxford University Press.

– 2009. *Moody Minds Distempered: Essays on Melancholy and Depression*. Oxford: Oxford University Press.

Rose, Deborah Bird. 2013. "In the Shadow of All This Death." In *Animal Death*, edited by Jay Johnston and Fiona Probyn-Rapsey, 1–20. Sydney: Sydney University Press.

Sandilands, Catriona. 2016. "Lavender's Green? Redux." In *Imperceptibly and Slowly Opening*, edited by Caroline Picard, 236–45. New York: The Green Lantern Press.

Styron, William. 1990. *Darkness Visible: A Memoir of Madness*. New York: Vintage.

Smith, Jeffery. 1999. *Where the Roots Reach for Water: A Personal and Natural History of Depression*. New York: North Point Press.

Solomon, Andrew. 2001. *The Noonday Demon: An Atlas of Depression*. New York: Scribner.

Williams, Raymond. 1973. *The Country and the City*. London: Chatto and Windus.

# 7 Climate Change as the Work of Mourning

ASHLEE CUNSOLO

## Introduction: Lament for the Land[1]

When I was five, a pond and thicket area down the street from my house was filled in and levelled while I was away. I remember coming home and finding my beloved ecosystem denuded of all greenery, and completely empty of the beavers and their dam, the minnows, the birds, and the countless rabbits and squirrels that had been a comforting and valued presence. I was devastated, consumed and overcome by grief and loss. I did not want to eat, or play, or go to school. I felt as though I had lost something deeply important, and intimately a part of the fabric of my life. It was the first time in my short life that I had become aware of the fragility of life – mine and others – and, from that moment, I found myself in a different life-world full of the awareness of the potential for death and injury to befall plants, animals, and ecosystems, aware of the corporeal acuteness of grief and mourning that could emerge from environmental destruction and degradation.

This experience was also the first in many moments of environmentally based grief – grief for the loss of non-human bodies, spaces, and places: the clear-cutting of a favourite hiking spot in British Columbia; the shooting of a mother black bear that I used to watch with her cubs every morning; the housing complex that disrupted a cougar corridor in Alberta; the dam that blocked salmon spawning near my home in the mountains; destruction of fertile farmland in Ontario; degradation of beloved ecosystems due to changes in climate; and grief that comes from witnessing the environmentally based mourning of friends and loved ones. I went through processes of grief and mourning for creatures and areas that were not human, but still caused significant feelings of loss within me.[2] It was also the first loss of many – human and non-human – that I have experienced, and those early days of grief and sadness created the foundations for my personal acts and responses to mourning – acts and responses that grew and transformed with each subsequent loss, each grief process, each work of mourning I undertook.[3] While each loss was experienced differently, uniquely, there was within each a memory a fleeting sense of that first death and of that early corporeal response to mourning an ecological loss.

These ecological grief experiences are certainly not unique to my personal experience; there are numerous people around the globe who have experienced or are currently experiencing grief and mourning responses to changes in their environment or due to the deaths of non-human entities, or understand the need to grieve for non-humans. Mental, emotional, and corporeal felt responses to environmental degradation and destruction have also been documented in response to severe drought (Albrecht et al. 2007; Berry et al. 2010; Speldewinde 2009; Berry et al. 2011), industrial activity and toxic exposure (Downey and Van Willigen 2005; Bevc et al. 2007), and localized ecological disasters such as hurricanes and oil spills (Palinkas et al. 1993; Havenaar et al. 2002; Picou and Hudson 2010). Despite the commonality of experiencing negative or emotional responses to environmental degradation, ecological losses do not appear in broader public and academic discourses concerning climate change – as though animal, vegetal, and mineral bodies are somehow constituted to be ungrievable in these broader narratives. Judith Butler (2004, xvi) expressed this unequal allocation of grievability well: "Some lives are grievable," she wrote, "and others are not; the differential allocation of grievability that decides what kind of subject is and must be grieved, and which kind of subject must not, operates to produce and maintain certain exclusionary conceptions of ... what counts as a livable life and a grievable death."

There are, tragically, bodies that do not matter in the public sphere, or bodies that have been disproportionately derealized from ethical and political consideration in global discourse: women, racial minorities, sexual minorities, peoples of different religions, certain ethnic groups, economically and politically marginalized groups, and Indigenous peoples, to name a few. To this list of derealized bodies I would add other-than-human bodies – animal, vegetal, and mineral. These bodies are experiencing the continued assaults of human activities, including myriad impacts from anthropogenic climate change. Yet these derealized bodies often "cannot be mourned because they are always already lost or, rather, never 'were,' and they must be killed, since they seem to live on, stubbornly, in this state of deadness" (Butler 2004, 33), at once alive but discounted. Yet mourn them people do, and mourn them we must.

This differential allocation of grievability became personally clear in 2006 when I began working with Inuit communities in the Canadian North on issues of climate change. Inuit in Canada are intimately connected to and reliant on their homeland, as the land ice and sea ice are the basis for their livelihoods, culture, and survival. The land is also an animate being with *whom* Inuit feel relational ties (the land is very often equated with the same language as people – i.e. whom rather than that). Indeed, my Inuit colleagues and friends in Northern Labrador have described the land as a close intimate, a mother figure and spiritual entity capable of response and reciprocity (see Cunsolo Willox et al. 2012, 2013a, b).

In the last decade, Inuit across Canada's North have been experiencing rapid changes in weather, water, snow, ice, wildlife, and vegetation, due to climatic and environmental change, and the resulting alterations in social and cultural activities, livelihoods, and land-based activities (Krupnik and Jolly 2002; Ford et al. 2006; Ford et al. 2008; Furgal 2008; Ford and Furgal 2009; Prowse and Furgal 2009; Ford et al. 2010; Cunsolo Willox et al. 2012, 2013a, b). Since Inuit lives and livelihoods are intimately intertwined with and reliant on the land, even subtle changes in climate and weather can cause significant environmental impacts – impacts which not only affect daily activities, but also cause strong emotional and mental responses (Norgaard 2006; Albrecht 2007; Fritze et al. 2008; Albrecht 2010; Berry et al. 2010; Berry 2011; Doherty and Clayton 2011; Norgaard 2011; Swim et al. 2011; Cunsolo Willox et al. 2012, 2013a, b). My colleagues and friends also frequently remarked that the current changes in climate and environment caused anxiety, fear, stress, worry, and anger as well as intense feelings of sadness, disorientation, grief, loss, and lament for a rapidly changing land. They also expressed place-based mourning and a sense of developing loss of a lifestyle, for a changed land, and for the affected plants and animals (Cunsolo Willox et al. 2012, 2013a, b).

Many people also shared a sense of anticipatory grieving for losses expected to come, but not yet arrived. Based on the rapidity of the changes in the region and the realization that these changes will not only continue, but will most likely worsen in severity and impact, many people indicated they were already imagining future losses, already experiencing levels of pain over what may come. In addition, as the community was engaged in anticipating the continuation of a changing climate at an increasing rate – and therefore of escalating disruption to and loss to the environment and non-human bodies – there was the associated memory and felt pain of previous loss and the anticipation of what future losses may feel like in comparison to these other losses. This anticipatory memory of loss is a mourning that begins before the break event, but is based in an understanding of the experience of other losses – land-based and not. That is, people are transferring their previous experiences and responses to grief and trauma from other situations and to varying degrees to their current and expected experiences with climatic and environmental change and the understanding of the intimate impact the environmental losses will have (see Albrecht et al. 2007; Albrecht 2010; Doherty and Clayton 2011; Cunsolo Willox et al. 2013b).

Despite these intense feelings and experiences, the grief and mourning experienced by individuals and communities globally to climate change seems strangely silenced in public climate change discourse. Indeed, the environment and non-human bodies do not normally appear within news media reports, dominant political discourses,[4] and even academic literature on climate change as something mournable or as a source of grief. This is a serious gap in academic literature, political practice, and media discourse around climate change, and does not match the lived experiences of many people around the globe who are already experiencing grief and mourning from environmental loss. Given the current global crisis of climate change, reconciling the private responses of environmentally based loss with the relative absence of this grief in public and academic spheres is of the utmost importance. Going further, grief and mourning have the unique potential to expand and transform the discursive spaces around climate change to include not only the lives of people who are grieving because of the changes, but also to value what is being altered, degraded, and harmed as something mournable. How can this recognition and reconstitution of non-human bodies as grievable within public and academic climate change discourse and literature occur? What can the work of mourning contribute to climate change discourse?

One possible avenue for reconstituting non-human bodies as grievable within the climate change arena is through the incorporation of the work of mourning in research, practice, public discourse, and action. By integrating the loss of non-human bodies and processes and the work of mourning into the climate change discourse, this work intends to create discursive and political space for the lived experience of climate-related grief and mourning and argue for thinking climate change with and through the work and labours of mourning. In so doing, this article extends the concept of a mournable body beyond the human in order to frame climate change as the work of mourning, and to discover what type of work this would be. It also examines the ethical and political implications that may emerge from thinking and acting with climate change as the work of mourning, and the ways in which this work can attend to both human bodies and non-human bodies through this type of environmental-based grief work. While mourning nature and non-humans is not new for the lived experiences of many people, what is new in this article is the extension of the work of mourning to the climate change discourse to discover the political and ethical possibilities emergent from uniting the work of mourning with climate change discourse – possibilities that can translate to public action, discursive shifts, and the enhancement of research, policy, and adaptation strategies. This work asks the question: can the work of mourning and the associated processes of grief, then, become a potent resource for politics or a grounding for ethics within the context of climate change?

## Ethical and Political Implications of Mourning

As we have read in the introduction to this collection, as well as in many chapters within, mourning is never strictly theoretical. It is real, it is work, and it binds us together with others (Butler 2004; Engle 2007). It is always already a condition of corporeality, and it is not something from which we can escape. It is affectively contagious, easily shared, and exposes the primacy of relational mental, emotional, and bodily ties. And it is a task that calls to us all through the relations we share with other bodies. It is both individualizing and unifying. And it is not just an act reserved only for those we know. Indeed, we can mourn for those whom we do not know, for those whom we will not know: the bodies lost in wars and acts of terrors (most recently: 9/11, Afghanistan, Iraq), natural disasters (Hurricane Katrina, the 2004 tsunami, the 2010 earthquake in Haiti, the 2011 earthquake in Japan), and humanitarian tragedies (the current drought in Somalia, deaths from poverty and disease).

We can also mourn for environmental degradation and destruction: the destruction of forests and farmlands; the devastation of landscapes for open-pit mining; the scarring of lands from tar sands projects; the leveling of mountain tops for mining; the pollution of rivers and lakes; the deaths of other creatures (beached whales, birds stuck in oil slicks, mass fish die-off, and animals struck by vehicles, to name a few); the melting of ice caps; the permanent loss of biodiversity through human-induced extinction; and the changes in lands all over the world because of climatic shifts and variability. These mourning responses can emerge through direct exposure to the actual event, or they can be mediated through news clips, stories from others, photographs, works of art, texts, video, or social media (Reser and Swim 2011). The ethical and political opportunities of the work of mourning arise in the myriad ways we choose to respond to and reciprocate – to others both human and non-human – and in the choices we make. But how can mourning do this? Does it not leave us immobilized through pain and suffering? Or as Butler (2004, 30) suggested: "Is there something to be gained from grieving, from tarrying with grief, from remaining exposed to its unbearability and not endeavouring to seek a resolution for grief through violence? Is there something to be gained in the political domain by maintaining grief as part of the framework within which we think our international ties? If we stay with the sense of loss, are we left feeling only passive and powerless, as some might fear? Or are we, rather, returned to a sense of human vulnerability, to our collective responsibility for the physical lives of one another?" While grief and mourning may indeed be unbearable, expose our very vulnerability, and at times make us feel as though we are powerless, it can also be understood "as the slow process by which we develop a point of identification with suffering itself" (ibid.). From this perspective, grief and mourning have the ability to mobilize, to galvanize, and to cause conscious action, not to privatize, silence, and subdue (Engle 2007), as we encounter the suffering and vulnerability of others and, in so doing, confront our own suffering and vulnerability. Mourning is also different than other corporeal or affective experiences that may occur because of loss, such as anger or rage. In contrast, mourning not only highlights collective vulnerability, it also commands that one respond through grief and pain, rather than solely rage or anger. As Butler (2004, xix) wrote: "And though for some, mourning can only be resolved through violence, it seems clear that violence only brings on more loss, and the failure to heed the claim of precarious life only leads, again and again, to the dry grief of an endless political rage."

What if we are expected not to mourn? What if we are asked to publicly shelve or bracket our mourning for something or someone or somewhere, as we have been asked to do with the impacts of climate change? What do we do when what could be mourned is stripped of its capacity to count as a grievable body in public discourse? Within the climate change discourse, the themes of grief and mourning do not explicitly emerge in discussions. While there is certainly coverage of the many current and possible impacts of climate change on animals and vegetation (perhaps most famously the depiction of the plight of the polar bears and the ubiquitous imagery of polar bears stranded on small ice floes), and while Indigenous populations around the globe – particularly those living in the Circumpolar North and in Small Island States – have been very vocal about the impacts of climate change on their lives, livelihoods, and local landscapes, grief and mourning are not highlighted within media, policy, or academic discourse, particularly grief centred around the loss of non-humans or ecosystems.

Despite the absence of mourning in climate change discourse, and the implicit framing of non-human bodies as non-grievable subjects, it is imperative to also highlight and share these grief experiences. Indeed, public mourning can be an important mechanism for political mobilization, the counteraction of dominant discourses around the derealization of non-human bodies, and for sharing the grief experienced from climatic and environmental change. For example, at the 2009 15th Conference of the Parties (COP 15) climate change negotiations in Copenhagen, the Tuvalu Delegation publicly shared their grief, sadness, and distress about the destruction of their coastlines and the rapid disappearance of parts of their island due to rising water levels. Ian Fry, one of the lead negotiators for Tuvalu, wept during his public speech representing Tuvalu's position, and this emotional outpouring of grief in a largely scientific and political setting served to disrupt the conversations momentarily, and to cause discomfort throughout the delegation (Farbotko and McGregor 2010 for an analysis of the impact of this moment on the negotiation process). Despite this event, to date, emotions such as grief and loss and the emotional impacts of climate change remain almost completely unexplored in climate change studies (ibid.; Norgaard 2011; Cunsolo Willox et al. 2012, 2013a, b).

To further illustrate the political potential of mourning, albeit from a very different literature and lived experience, I draw on the concerted, conscious political mobilization movement that reconstituted the AIDS body as mournable. Although the first medical reports of AIDS appeared in North America in 1981, for many years AIDS bodies were marginalized within public discourse. Those living with AIDS, as well as their loved ones, were derealized from the public sphere of mourning in many ways (Butler 1993).[5] The re-constitution of the AIDS body as something grievable required significant theoretical, political, and cultural activism and re-codification, and countless individuals uniting together to attempt to redefine the AIDS body as something mournable and something absolutely imperative to grieve publicly and openly. This process would not have succeeded without the conscious creation of public acts of mourning: public testimonies and eulogies, elaborate funerals, public memorials, the creation of the AIDS quilt (www.aidsquilt.org), the famous photograph of AIDS activist David Kirby at the end of his life (photographed by Therese Frare[6]), plays and theatrical productions, and even Hollywood films helped to reconstitute the AIDS body as a human body in broader social discourse (Butler 1993), vulnerable like our own, whose suffering and destruction is tragic, grievable, and an appropriate source of mourning. That is, with concerted theoretical, political, social, and cultural activism and reframing, and through public outpourings and testaments of grief, previously marginalized and ungrievable bodies became socially constituted to again be mournable in public, political, and ethical discourses. In this movement, grief and the work of mourning became a driving factor and a potent political strategy to break through the marginalization to reconstitute the AIDS body as something worthy of and appropriate for mourning in the public discourse.

While I hasten to add that this example does not intend in any way to compare the deaths of those who passed from AIDS or the mourning of their loved one with the deaths of non-humans from climate change, nor to ignore or conflate the politics of sex, race, gender, and marginalization within the AIDS movement and literature with the politics of climate change, there are some very important lessons from the hard work, dedication, political action, protest, and organization of the AIDS movement that can potentially assist in the climate change discourse. If we map this political and ethical potential for discursive transformation that emerges from the example of mourning and the AIDS movement onto the climate change discourse, the potential for expanding environmental politics and ethics through the work of mourning emerges – a politics and ethics that may expose the inherent injustice in silenced deaths (Spargo 2004) and may counteract the derealization of nature, and those who mourn it, in dominant climate change discourses.

## Climate Change as the Work of Mourning

What, then, can be learned from this political change in the AIDS movement that can extend to climate change? What are the possible mechanisms of mourning that are productive for understanding the impacts of climate change as grief work? First, environmentally based grief needs to continue to be spoken loudly and often, in both private and public settings. Just as Ian Fry and the Tuvalu Delegation did in Copenhagen in 2009, and just as many Indigenous peoples, artists, photographers, and writers continue to do through their stories and visual media, this grief for non-human bodies and processes – particularly grief experienced through changes in climate and environment – needs to be shared broadly to counteract the violence of derealization to repopulate the climate change discourse with the voices and experiences of environmentally based mourning, and to socially constitute other-than-humans as something mournable and grievable. While this climate-related grief and mourning is emerging around the globe from people living at the frontlines of climate change – peoples in the Circumpolar region, Small Island States, and Australian farmers, to name a few (Albrecht et al. 2007; Speldewinde et al. 2009; Berry et al. 2010; Farbotko and McGregor 2010; Berry et al. 2011; Cunsolo Willox et al. 2012, 2013a, b) – this is a work for us *all*. We need to continue to eulogize and read out the names of those non-humans that have been lost, or are close to disappearance. We need to continue to speak the names of the extinct (or close to) at public events, in classrooms, and in private settings. We need to continue to create works of art, literature, and writing extolling this environmentally based grief and loss. An interesting example of this is the creation of the Mass Extinction Memorial Observatory (MEMO) currently being erected in the United Kingdom on the Isle of Portland, which will host carvings of all the plants and animals that have become extinct in modern times. The MEMO is also meant to celebrate biodiversity and the importance of all creatures on this planet, and every year on 22 May for International Biodiversity Day, a bell will toll marking all animals and plant species who have passed (www.memoproject.org).

Second, this grief needs to be witnessed and shared by others, whether they have experienced environmentally based grief due to climate change or not. This shared witnessing allows the opportunity for individuals to connect with shared responsibility for this grief from a global process, and to understand this mourning as personal, political, and ethical – as illustrative of the injustices perpetuated against the other-than-human world by human actions and illustrative of the injustices experienced by those who currently bear the burden of this type of mourning. We need to continue to create works of art, academic treatises, policy statements, memorials, monuments, theatre productions, and public forums, like the Mass Extinction Memorial Observatory, to share this ecological grief and to provide places for people to go and collectively mourn – mourn in ways that are indeterminate, adaptive, open-ended, shared, and premised on a submission to transformation.

Third, the work of mourning further exposes our individual and collective vulnerability not only to other humans who are currently experiencing the burden of global climatic changes, but also to non-human bodies and processes transforming because of climate change. This shared vulnerability, emergent from understanding climate change as the work of mourning, can extend beyond the human to be more inclusive of all bodies affected and being affected. This mutual vulnerability may also be a powerful mechanism to incite public participation in ecological-grief-related events mentioned in the previous paragraph, and subsequently to enhance adaptation and resilience through shared grief, collective mourning, and group action. In addition, this shared vulnerability may itself be a mechanism for shared resilience to change, as people have the opportunity to share their grief, take comfort in communities formed in response to climate-related mourning, and come together in unity to effect change.[7]

Fourth, reframing a movement such as climate change through mourning can repopulate the literature and discourse by including emotions and grief as meaningful and powerful aspects of climate change, and can recognize publicly the substantial impacts of climate change and those who mourn the changes. Grief also offers a counter-narrative or alter-narrative to those highlighting the problems of climate change for humanity and those focusing on adaptation and mitigation (Randall 2009). While both of these narrative streams possess very important information and perspectives, they can be deepened through the inclusion of grief and opportunities for public mourning. In so doing, further opportunities for enhancing resilience and adaptive capacities through productive and shared mourning and public acts of grief may also emerge.

Although mourning can lead to feelings such as anger, rage, or hatred, if mourned with intent to grieve and respect what was lost and to heal, mourning has the capacity to be a more psychologically healthy emotion to incite political action (rather than action premised on rage or hatred). Sharing in mourning and working through the grief process may assist in psychological resilience to the changes (Randall 2009; Cunsolo Willox et al. 2013b) as well as furnish a sense of political and ethical community in response to the changes based on something beyond violence or rage (Butler 2004), yet still understands the place for anger and rage to emerge against the injustice of the deaths of other species and creatures due to anthropogenic climate change and human action. Mourning works to move beyond violence and hatred, towards a place of uniting through the creation of public memorials, art, theatre, protests, and the sharing of grief – actions which may also enhance resilience to climate change. Furthermore, without speaking about loss and without participating in the work of mourning, psychological change, transformation, and resilience may be hindered (Randall 2009).

This list is not meant to be limiting or exhaustive, but rather, a starting point for examining opportunities to cohesively unify and engage with climate change issues through shared global grief and mourning for what has been, currently is, and will be lost in the other-than-human worlds. This list does not negate the artists, writers, academics, and citizens who are creating works of art and thought with nature at their core, monuments or memorials to commemorate the destruction of other species and creatures, and creations to share distress and sadness. Instead I mean to say that we not only need more of that, we also need to actively seek opportunities for unification through the grief and mourning associated with climate change on larger political, geographic, and ethical scales – unities and projects that are less fragmented and that can become part of dominant discourses about climate change-related grief and mourning. Furthermore, mourning has the unique capacity to transcend boundaries for shared understanding, empathy, ethics, and politics, and to act as a starting point to encourage more direct engagement with environmentally based grief and mourning and the potent ethico-political opportunities that emerge through shared and mobilized mourning.

By conceptualizing climate change as the work of mourning, a space opens up for land-based grief and the loss experienced as a result of climate change to be expressed, shared, and discussed. It publicizes what has previously been pushed into the margins of the private sphere, and emphasizes the intimate and transcorporeal connections shared body to body across species and boundaries and spatial and temporal scales. Thinking climate change as the work of mourning is thus a powerful ethical and political act. To mourn shows respect for and connection to what is mourned – in this case, non-human bodies, processes, and ecosystems – and to reconstitute bodies that have been lost in their individuality and importance. Through the mourning process bodies experience the transformative power of loss that Butler identified and are open to the capacity to be moved, altered, and affected by the loss of bodies beyond humans.

This framing of climate change as the work of mourning also exposes the weaknesses in our theoretical constructs and discursive framing, as the lived experiences of people living with and through environmental grief, place-based mourning, and the loss of the land and non-human entities have outstripped what is conceptualized in the theoretical work of mourning. Furthermore, the problem is not only with our conception of nature, or the socially constituted notions of what counts as mournable or grievable – it is also with our conception of those who grieve for the ungrievable, such as Indigenous groups, farmers, or those who rely closely on the natural environment. And although the media and academic studies highlight the negative environmental impacts of climate change (loss of species, glacier recession, polar ice cap melt, coral reef bleaching and destruction), and the subsequent impact on people (loss of livelihoods, migration, relocation), grief and mourning as an explicit point of focus or discussion is still absent. While people may respond to the news reports or research with their own grief, it has not been a direct or overt point of study or dialogue. Going further, perhaps mourning non-human entities has not received much public attention because those most likely to partake in this work in response to climate change are themselves bodies that do not usually matter within policy and discourses; those who are most likely to grieve the loss of the land and climatic and environmental degradation are precisely those who are themselves most often marginalized. This marginalization also means that the vulnerabilities these groups have to climatic and environmental change, and the resulting emotional responses such as loss and grief, are often ignored or absent in larger public discourse.[8]

This is where mourning, and the subsequent work, becomes a productive and politically and ethically salient method for framing the impacts of climate change on people, populations, and ecosystems, and provides opportunities for changing the discourse of climate change and mourning to include non-human bodies and processes. Just as the public work and acts of mourning of AIDS sufferers helped to realize AIDS bodies in public discourse and highlight their vulnerability, so too can public works and acts of mourning of non-human bodies help to reconstitute these bodies as real, mournable, and grievable. The work of mourning highlights vulnerability to loss and to change experienced by other people on this planet dealing with climate change – a vulnerability that we all share as lived bodies. By publicizing this land-based mourning, we make this ecological grieving and this vulnerability to climatic and environmental change visible to the dominant discourse and assist in the re-constitution of non-human bodies, and those who mourn them, as bodies that matter, and bodies count in the work of mourning. For example, there are numerous ways in which public testimonials and the witnessing of grief and mourning can lead to healing, closure, and justice for those who have suffered trauma and atrocities: the Canadian Truth and Reconciliation commissions in response to residential schools, widespread abuse, forced assimilation, and the killing of sled dogs; the Nuremburg and other Second World War tribunals; and the International Criminal Tribunals for Rwanda and Bosnia (to name a few examples).

This sense of healing and community building and recognition of justice experienced through these public gatherings and political movements may also hold the same healing and political unification within the context of climate change-related grief and mourning. To move forward in global ethical and political dialogues around climate change, we must look for opportunities to make public the grief and mourning experienced due to changes in the land and ecosystems. Climate change-related grief, and the subsequent labours of mourning, disrupt the marginalization and derealization of bodies (human and non), and can act as ethical and political resistance to the discourses that socially constitute non-human bodies as unmournable. We are all vulnerable to climate change, and we are all vulnerable to death and loss from climate change. Thinking climate change *as* the work of mourning provides the opportunity to learn from the deaths, or the potential deaths, of bodies beyond our own, and beyond our species to unite in individual and global action and response.

Mourning is work for us *all*. Climate change is also work for us all, as citizens of this planet, and as those who hold the responsibility for the changes and for the changes perpetuated on our feathered, furred, scaled, insect, microbial, and plant kin. Climate change represents the largest human-induced global ecological threat experienced on this planet, but there is still much fragmentation in research, politics, policy, and action. Mourning may be one such mechanism to assist with finding common ground among peoples from different countries, cultures, and climates to unite together, share experiences, and creatively enhance resilience and adaptive capacities. Indeed, further research and work will benefit from including stories and reports of ecologically based grief and mourning experienced by peoples in regions where the most serious impacts of climate change are being felt, such as the Polar regions, low-lying island states, and ecologically sensitive ecosystems, as well as by people whose grief and mourning are in response to mediated images and texts (Reser and Swim 2011). In addition, research examining culturally based patterns or responses to guilt, and an analysis of the ways in which previous traumas (ecologically or otherwise) were experienced, and comparing, contrasting, and mapping onto the grief experienced through climate change is also another important area for further study. Finally, there is also the potential for fruitful and fecund research by examining the intersection of grief and guilt within the context of climate change. Often we grieve and mourn for that which we have no control or part; within the context of anthropogenic climate change, however, the changes experienced throughout the globe, and the impacts on humans and non-humans alike, are directly related to human actions, and thus although we may mourn, we are also implicated in our actions. This tension between mourning what has been lost or what is changing coupled with guilt over our own actions that have led to these changes is an important area for further research and consideration (there may be helpful grief-related literature examining survivor guilt or coping mechanisms from those who have caused accidental deaths).

## Concluding Thoughts

As has been argued through this paper, mourning, and the associated work, is one of the most fundamental capacities of being human, and may provide the means to move ever deeper into the sensorial present with humans and non-humans. As Nass (2008, 170) wrote, the work of mourning "opens up the possibility of a social or political space to accommodate all others." Thinking climate change as the work of mourning means that we are ethically and politically implicated not only by what is happening to animal, vegetal, and mineral bodies and assemblages, but also in the choice to respond. This work, then, opens up the opportunity to mourn each and every "body" differently and publicly, rather than as an aggregation under an abstract concept or as a lived experience mourned privately, silenced in public and academic narratives of mourning, or simply not mourned at all. This ecological grief work, and the resulting ethics, offers something for us all to learn in the new global reality of climate change. How we reply may be different, and our responses may not always be up to the task, but the ecological work of mourning "is hope, the hope for unimaginably better futures for unknown and unknowable recipients in a space left to them" (Houle 2007, 163). In this ecological work of mourning, in our individual and collective grief, and in the possibilities for transforming our political and ethical landscapes that climate change is offering we may, as Houle (ibid., 164) wrote,

> glimpse a unique constellation of human withness, of immanent multiplicity: what is always everywhere asking for hospitality just where we are not yet ready for it. There was, or perhaps there is calling here. A calling for a unique form of response: what might come forth in the wake of attending to these sorts of deaths? Perhaps the featureless, nameless Face of the democracy to come: An ecological democracy-to-come that includes animal, vegetal, and mineral bodies and ecosystems within the work and labours of mourning – a new ethical and political future in response to environmentally-based grief and lament, and to counteract the destruction and degradation of our non-human kin.

This grief also needs to be shared, for as Derrida (2001) explained in mourning "speaking is impossible, but so too would be silence or absence or a refusal to share one's sadness" (72). Mourning, then, is about sharing one's sadness and bearing witness through our own lives and bodies to the lives, bodies, objects, ideals, and abstractions that have come before us and of which we have lost. I, myself, have stood up and testified for human loved ones. I have publicly shared my grief. I have publicly mourned. I have written and spoken and expressed my grief and participated in the work of mourning for human intimates. But I have not done this work publicly for non-humans, for the loss of beloved ecosystems and the destruction of animal, plant, and mineral kin, for the affective grief and mourning I feel when confronted with friends' and colleagues' environmentally based grief and mourning. Perhaps then this chapter is also, in a way, an expression of grief for the loss and pain and mourning that I feel for the derealization of non-humans; an environmental eulogy for the destruction and violence perpetuated on non-human bodies, for the anticipatory grief I feel for the future changes, and for the empathetic sadness I feel through the pain that my friends and colleagues experience because of changing climate and environments. This paper, then, is but a small step towards my environmental grief work, my ecological work of mourning.

Recasting climate change as the work of mourning means that we can share our losses, and to encounter them as opportunities for productive and important work to be given primacy and taken seriously. It also provides the opportunity to stand up and publicly object to injustice (Spargo 2004): injustice to non-human bodies; injustice to Indigenous bodies; injustice to the bodies that have been derealized and socially constituted as unmournable. The work of mourning brings back these bodies to the foreground as something worthy to be mourned through productive, transformative, interminable, and never-ending work – work to be conducted and taken up, right now, before our death and the death of others; work that may allow for a deeper understanding of our relationships with other bodies, human and non-human – a new ecological ethic and platform for unification and action premised upon and mobilized through the work and labours of mourning.

## NOTES

1 This chapter originally appeared in *Ethics and the Environment* and has been reproduced with permission in modified form for this collection. This article was also the original inspiration for the collection and, due to its reception when published, indicated the need for a work that compiled some of the leading thinkers in this area together. Some of the original pieces of this article have been moved into the introduction for this collection.

2 It is important to note that I am not speaking here of companion animals or pets, but rather, of non-human entities, such as animal, vegetal, and mineral bodies and ecosystems, which are not usually acknowledged to cause substantial feelings of loss.

3 For the purposes of this paper, the work of mourning is defined as the personal, ethical, and political tasks and ways of living to which a body must

attend and in which a body must partake in the event, perceived eventuality, and possibility of loss. This work is always and simultaneously personal, political, and ethical, and corporeally embodied. It is an individualizing work, as loss is experienced differently by everyone, but simultaneously it is also a unifying work, bringing people together through collective experiences of sharing grief.

4 A potential attempt to bring non-humans to the centre of political discourse may be found in Bolivia's push to create an historic "Law of Mother Earth," which would grant nature the same rights and legal protections as humans. This law is set to recognize that nature, the environment, and all entities within, have a right to life and to existence – an existence free from pollution, degradation, and destruction. Bolivia is also in the process of creating a Ministry of Mother Earth to oversee the new legislation. See, for example, http://www.guardian.co.uk/environment/2011/apr/10/bolivia-enshrines-natural-worlds-rights.

5 Infamously, then-president Ronald Reagan did not acknowledge the disease publicly or in political discourse. Even after the death of friends and acquaintances of AIDS, such as Rock Hudson, he still did not mention the pandemic publicly until 1987 (and after almost 60,000 cases of AIDS were diagnosed in the United States alone).

6 To view this photograph, and the associated photographic documentary, *The Photo That Brought Aids Home* by Therese Frare, published online through *Life* magazine, please visit www.life.com/gallery/45701/the-photo-that-brought-aids-home#index/6.

7 For my own work with Inuit in Nunatsiavut, Labrador, climate change and the resultant emotional, mental, social, and cultural responses have provided an opportunity and a space for people to come together to talk about the issues facing their communities and ways to adapt and strategies for keeping communities healthy. In particular, having the opportunity to talk about the grief, loss, and mourning experienced in response to these changes was not only healing and therapeutic, but also mobilizing and empowering (Cunsolo Willox 2012, 2013a, b; cf. Fritze et al. 2008; Berry 2009).

8 It is important to emphasize that here I am speaking about the absence of grief and mourning within media reports, policy documents, and academic discourse. There are indeed numerous articles published on the human dimensions of climate change within an Indigenous and/or socio-economically marginalized context. For example, within a Canadian context and as of December 2010, there was approximately one scientific study published per 1,000 Inuit on the human dimensions of climate change (and approximately one per 450 Inuit published on climate change-related issues in general) (Ford et al. 2012). While these studies cover a wide range

of topics, grief and mourning are relatively absent from the literature (see n. 6 for a discussion of a literature search on this topic).

## REFERENCES

Albrecht, Glenn. 2010. "Solastalgia and the Creation of New Ways of Living." In *Nature and Culture: Rebuilding Lost Connections*, edited by Sarah Pilgrim and Jules Pretty, 217–34. London: Earthscan.

Albrecht, Glenn, Gina-Maree Sartore, Linda Connor, Nick Higginbotham, Sonia Freeman, Brian Kelly, Helen Stain, Anne Tonna, and Georgia Pollard. 2007. "Solastalgia: The Distress Caused by Environmental Change." *Australian Psychologist* 15: S95–8.

Berry, Helen Louise. 2009. "The Pearl in the Oyster: Climate Change as a Mental Health Opportunity." *Australasian Psychiatry* 17, no. 6: 453–6.

Berry, Helen Louise, Kathryn Bowen, and Tord Kjellstrom. 2010. "Climate Change and Mental Health: A Causal Pathways Framework." *International Journal of Public Health* 55: 123–32.

Berry, Helen Louise, Anthony Hogan, Jennifer Owen, Debra Rickwood, and Lyn Fragar. 2011. "Climate Change and Farmer's Mental Health: Risks and Responses." *Asia-Pacific Journal of Public Health* 23, no. 2: 1195–1325.

Bevc, Christine A., Brent K. Marshall, and J. Steven Picou. 2007. "Environmental Justice and Toxic Exposure: Toward a Spatial Model of Physical Health and Psychological Well-Being." *Social Science Research* 36: 48–67.

Brault, Pascale-Anne, and Michael Naas. 2001. "To Reckon with the Dead: Jacques Derrida's Politics of Mourning." In *The Work of Mourning*, edited by Pascale-Anne Brault and Michael Naas, 1–30. Chicago: University of Chicago Press.

Butler, Judith. 1993. "Critically Queer." *GLQ: A Journal of Lesbian and Gay Studies* 1, no. 1: 17–32.

– 2004. *Precarious Life: The Powers of Mourning and Violence*. New York: Verso.

Cunsolo Willox, Ashlee, Sherilee Harper, Victoria Edge, Karen Landman, Karen Houle, James D. Ford, and the Rigolet Inuit Community Government. 2013. "'The Land Enriches the Soul:' On Environmental Change, Affect, and Emotional Health and Well-Being in Nunatsiavut, Canada." *Emotion, Space, and Society* 6: 14–24.

Cunsolo Willox, Ashlee, Sherilee Harper, James D. Ford, Victoria Edge, Karen Landman, Karen Houle, Sarah Blake, and Charlotte Wolfrey. 2013. "Climate Change and Mental Health: An Exploratory Case Study from Rigolet, Nunatsiavut, Labrador." *Climatic Change* 121, no. 2: 255–70.

Cunsolo Willox, Ashlee, Sherilee Harper, James D. Ford, Karen Landman, Karen Houle, Victoria Edge, and the Rigolet Inuit Community Government. 2012.

"'From this Place and of this Place': Climate Change, Health, and Place in Rigolet, Nunatsiavut, Canada." *Social Sciences and Medicine* 75, no. 3: 538–47.

Dean, John G., and Helen J. Stain. 2010. "Mental Health Impact for Adolescents Living with Prolonged Drought." *The Australian Journal of Rural Health* 18: 32–7.

Derrida, Jacques. 2001. "Paul de Man, December 6, 1919–December 21, 1983." In *The Work of Mourning*, edited by Pascale-Anne Brault and Michael Naas, 69–75. Chicago: University of Chicago Press.

Doherty, Thomas, and Susan Clayton. 2011. "The Psychological Impacts of Global Climate Change." *American Psychologist* 6, no. 4: 265–76.

Downey, Liam, and Marieke Van Willigen. 2005. "Environmental Stressors: The Mental Health Impacts of Living Near Industrial Activity." *Journal of Health and Social Behavior* 46, no. 3: 289–305.

Engle, Karen. 2007. "Putting Mourning to Work: Making Sense of 9/11." *Theory, Culture, and Society* 24, no. 1: 61–88.

Farbotko, Carol, and Helen McGregor. 2010. "Copenhagen, Climate Science, and the Emotional Geographies of Climate Change." *Australian Geographer* 41, no. 2: 159–66.

Ford, James, Lea Berrang-Ford, Malcolm King, and Chris Furgal. 2010. "Vulnerability of Aboriginal Health Systems in Canada to Climate Change." *Global Environmental Change* 20, no. 4: 668–80.

Ford, James, Kenyon Bolton, Jamal Shirley, Tristan Pearce, Martin Tremblay, and Michael Westlake. 2012. "Mapping Human Dimensions of Climate Change Research in the Canadian Arctic." *Ambio* 41, no. 8: 808–22.

Ford, James, and Chris Furgal. 2009. "Foreword to the Special Issue: Climate Change Impacts, Adaptation, and Vulnerability in the Arctic." *Polar Research* 28: 1–9.

Ford, James, Barry Smit, and Johanna Wandel. 2006. "Vulnerability to Climate Change in the Arctic: A Case Study from Arctic Bay Canada." *Global Environmental Change* 16: 145–60.

Ford, James, Barry Smit, Johanna Wandel, Mishak Allurut, Kik Shappa, Harry Ittusarjuats, and Kevin Qrunnuts. 2008. "Climate Change in the Arctic: Current and Future Vulnerability in Two Inuit Communities in Canada." *The Geographical Journal* 174, no. 1: 45–62.

Fritze, Jessica G., Grant A. Blashki, Susie Burke, and John Wiseman. 2008. "Hope, Despair and Transformation: Climate Change and the Promotion of Mental Health and Well-Being." *International Journal of Mental Health Systems* 2, no. 13: 13. doi:10.1186/1752-4458-2-13.

Furgal, Chris. 2008. "Climate Change Health Vulnerabilities in the North." In *Human Health in a Changing Climate: A Canadian Assessment of Vulnerability and Adaptive Capacity*, edited by Jacinthe Seguin, 303–66. Ottawa: Health Canada.

Havenaar, Johan M., Julie G. Cwikel, and Evelyn J. Bromet, eds. 2002. *Toxic Turmoil: Psychological and Societal Consequences of Ecological Disasters*. New York: Kluwer Academic/Plenum Publishers.

Houle, Karen. 2007. "Abortion as the Work of Mourning." *Symposium: Canadian Journal of Continental Philosophy* 11, no. 1: 141–66.

Krupnik, Igor, and Dyanna Jolly, eds. 2002. *The Earth Is Faster Now: Indigenous Observations of Arctic Change.* Fairbanks: Arctic Research Consortium of the United States.

Norgaard, Kari. 2011. *Living in Denial: Climate Change, Emotions, and Everyday Life*. Cambridge, MA: MIT Press.

– 2006. "'People Want to Protect Themselves a Little Bit': Emotions, Denial, and Social Movement Nonparticipation." *Sociological Inquiry* 76, no. 1: 372–96.

Palinkas, Lawrence A., John S. Petterson, John Russell, Michael A. Downs. 1993. "Community Patterns of Psychiatric Disorders after the Exxon Valdez Oil Spill." *The American Journal of Psychiatry* 150, no. 10: 1517–23.

Picou, J. Steven, and Kenneth Hudson. 2010. "Hurricane Katrina and Mental Health: A Research Note on Mississippi Gulf Coast Residents." *Sociological Inquiry* 80, no. 3: 513–24.

Prowse, Terrence, and Chris Furgal. 2009. "Northern Canada in a Changing Climate: Major Findings and Conclusions." *Ambio* 38: 290–2.

Randall, Rosemary. 2009. "Loss and Climate Change: The Cost of Parallel Narratives." *Ecopsychology* 1, no. 3: 118–29.

Reser, Joseph P., and Janet K. Swim. 2011. "Adapting to and Coping with the Threat and Impacts of Climate Change." *American Psychologist* 66, no. 4: 277–89.

Sartore, Gina Maree, Brian Kelly, Helen Stain, Glenn Albrecht, and Nick Higginbothom. 2008. "Control, Uncertainty, and Expectations for the Future: A Qualitative Study of the Impact of Drought on a Rural Australian Community." *Rural and Remote Health* 8: 950.

Spargo, R. Clifton. 2004. *The Ethics of Mourning.* Baltimore: The Johns Hopkins University Press.

Speldewinde, Peter, Angus Cook, Peter Davies, and Philip Weinstein. 2009. "A Relationship between Environmental Degradation and Mental Health in Rural Western Australia." *Health and Place* 15, no. 3: 880–7.

Thompson, Clive. 2008. "Global Mourning: How the Next Victim of Climate Change Will Be Our Minds." *Wired Magazine* 70 (January). http://www.wired.com/images/press/pdf/globalmourning.pdf. Accessed 21 January 2014.

# 8 Auguries of Elegy

## *The Art and Ethics of Ecological Grieving*

JESSICA MARION BARR

A Robin Red Breast in a Cage
Puts all Heaven in a Rage.

WILLIAM BLAKE, "AUGURIES OF INNOCENCE"

An elegiac mode is appropriate, given the loss of species, of habitats, of old forms of life – "old" here standing in for anything that happened earlier than last week ... Environmental language, however, speaks elegies for an incomplete process, elegies about events that have not yet (fully) happened ... They fuse elegy and prophecy, becoming elegies for the future.

TIMOTHY MORTON

## Introduction

In the trailer for artist Chris Jordan's film *Midway*, viewers are invited to "Come ... on a journey through the eye of beauty, across an ocean of grief ... and beyond" (2013) as he documents the trash-filled bellies of baby albatrosses, dead after being fed plastics from the Pacific "Gyre" by their unwitting parents. Jordan further states that "*Midway* explores the plight of the Laysan albatross plagued by the ingestion of our plastic trash. Both elegy and warning, the film explores the interconnectedness of species, with the albatross on Midway [Island] as a mirror of our humanity" (2013). Jordan is one of many contemporary ecologically engaged artists whose elegiac work explores an ethics of ecological melancholic grieving as a pedagogical and activist strategy. As an artist-researcher my work also engages with ecological elegy and what Patricia Rae calls "resistant mourning" or "activist melancholia" (2007, 19), revisiting the ethics of mourning that began with modernist writers and visual artists. As Timothy Morton (2010) states, elegy is an appropriate mode for expressing the complex feelings that may accompany an awareness of the state of human-caused climate change and the ailing health of the biosphere. In order to understand these contemporary concerns, I look back to the early part of the twentieth century, when the cataclysms of global wars and their destructive effects on human and more-than-human life inspired the first generation of modern "resistant" elegists. Like modernist anti-war elegies, the contemporary ecological (visual) elegies that I will examine engage in deliberate, ethical mourning and melancholia in an ultimately hopeful attempt to create a better future.

## Elegies

In *Modernism and Mourning*, Patricia Rae points to the relationship between the overwhelming disasters of the late twentieth and early twenty-first centuries (AIDS, genocides, terrorism, wars) and a corresponding explosion of "public and academic interest in how we mourn and in the question of whether there is social progress to be gained from experiences of loss" (2007, 13). In her view, this intense and "ongoing preoccupation with mourning provides a strong motive for revisiting the literature produced during and between the First and Second World Wars, because the 'work of mourning,' or, more precisely, the 'resistance' to this work, was central to this literature, shaping both its themes and its formal experiments" (ibid.). Rae defends the continuing relevance of these politicized "resistant" anti-war elegies, stating that they "leave mourning unresolved without endorsing evasion or repression; indeed, they portray the failure to confront or know exactly what has been lost as damaging. They encourage remembering where memory has been repressed, and they expose the social determinants for troublesome amnesia. At the same time, they resist the narratives and tropes that would bring grief through to catharsis, thus provoking questions about what caused the loss, or about the work that must be done before it is rightly overcome. They raise questions about the social forces that have prevented the work of mourning from being accomplished, and they offer alternatives to the consolatory strategies that have been widely deployed and that threaten to introduce a whole new round of loss and grieving" (ibid., 22–3). Rae suggests that, in its concern about preventing future losses, this type of resistant mourning "encourages work for positive social change" (ibid., 23).

Prior to the World Wars, elegies (Western European and British in particular) were premised on resolving mourning and finding consolation and comfort in nature's cycles, in the flourishing of the nation-state, and in visions of eternal life in heaven. As such, elegies (written, musical, and visual) aided in what Freud termed "the work of mourning." In his seminal essay "Mourning and Melancholia," Freud examines the "normal" compensatory, resolvable "work of mourning" (grieving and eventually "getting over" the death of a loved one by "working" to break the attachment of the mourner to the deceased and replace it with a new libidinal attachment to someone or something else) versus "pathological" melancholia (endless, irresolvable, depressive mourning that "behaves like an open wound" and refuses to sever from the lost object) (1957, 253). Traditional elegy, visually expressed in artworks such as French Classical painter Nicolas Poussin's *Et in Arcadia Ego* (c. 1650), aided in the process of "healthy" mourning, providing grievers with libidinal substitutes in nature, nation, religion, and so on.

But as the cataclysm of the First World War and then the incomprehensible horrors of the Spanish Civil War and the Second World War unfolded, many writers, activists, and artists began to realize that, when confronted with the traumatizing, unjust, and uncountable losses of the wars, the idea of resolvable and consolatory mourning was not only impossible, it was also unethical. Artistic and literary responses to these conflicts – especially retrospective ones – could only be anti-elegiac, or truly "modern" elegies of perpetual, "ethical" mourning. The comfort provided by conventional mourning and its artistic representations could lull people into passive forgetfulness, which could result in history repeating itself. Thus an insistence on remembering, on dwelling in the pain of unjust loss, was deemed necessary by politically engaged artists as a sort of ethical insurance that such losses would never be allowed to happen again.

Modern elegies are difficult and complicated. "Anything simpler or easier," Jahan Ramazani (1994) notes, "would betray the moral doubts, metaphysical skepticisms, and emotional tangles that beset the modern experience of mourning and of self-conscious efforts to render it" (x). A psychoanalytic model of compensatory, resolvable mourning is "inadequate for understanding the twentieth-century elegy" (ibid., xi) as well as being inadequate for comprehending or commemorating the horrors of the twentieth century's wars, mass deaths, myriad other atrocious injustices, and ecological catastrophes. Faced with the impossibility of traditional mourning and elegizing, artists like Virginia Woolf believed that "art must be stripped of compensatory literary tropes in order to soberly confront the horror and politics of manufactured death" (Clewell 2004, 214). Traditional elegies, both literary and visual, would not do. As a way of understanding post–First World War elegies, Ramazani proposes "the psychology of melancholia or melancholic mourning," arguing that these works tend "not to achieve but to resist consolation, not to override but to sustain anger, not to heal but to reopen the wounds of loss" (1994, xi). As such, "the elegy flourishes in the modern period by becoming anti-elegiac (in generic terms) and melancholic (in psychological terms)" (ibid.) In order to, as Margot Heinemann puts it in her Spanish Civil War elegy, "Grieve in a New Way for New Losses." Contrary to traditional hopes for a comforting "eternal life" through commemorative elegy, modern elegists force an ongoing confrontation with death. It is death, not life, that is eternal in both the literary and visual modern elegy.

Whereas modern poets use the traditional elegiac form as a basis for their experiments in radical poetic grieving (ibid., 1), modern pictorial elegists find, in some cases, radically new structures and imagery for their fraught visual elegies.[1] Kerr Eby's etching *St Mihiel, September 13, 1918: The Great Black Cloud* (1918), as well as John P.D. Smith's *Death Awed* (c. 1918–19, a disturbing depiction of the grim reaper on the battlefield, shocked by the sight of a pair of boots standing upright with their owner's splintered shin bones emerging from them) are, according to Elizabeth Helsinger, "lesser-known counterparts to the bitter poems of grief and protest by war poets like Siegfried Sassoon, Isaac Rosenberg, and Wilfred Owen. The etchers condense the poetry's extended

Figure 8.1 | Kerr Eby, *St Mihiel, September 13, 1918: The Great Black Cloud* (1918/1934). Collection of the Mattatuck Museum, Waterbury.

unfolding of experience into a single stark image ... They exploit the impersonality of the visual image – its inability to utter the first person 'I' – to stress the obliteration of the individual in time of war" (2010, 675). Sandra M. Gilbert (2006) describes elegy as "a genre disfigured by a traumatized modernity" (368), and pictorial responses to the World Wars and the Spanish Civil War are an especially apt representation of this disfigured poetics. In "Grieving Images: Elegy and the Visual Arts," Helsinger (2010) states that modern visual elegy "continually confronts us with the moment of overwhelming grief, the particular event of death that gestures to the enormity of mass destruction. When elegy begins to refuse the 'normal' work of mourning (the movement toward consolation), its pictorial forms acquire new power" (676). These pictorial elegies, in melancholically illustrating memory, act as a sort of "instrument of war" (Pablo Picasso, in Martin 2002, i), firing powerful visual warning shots to make "war against war" (to paraphrase the title of a 1923 print by

Käthe Kollwitz): to warn, to keep the wound open, not maliciously, but as an ethical salvo against complacency and forgetting. There is a particular potency in the visual image, one that Helsinger refers to as its "immediacy," which, she claims, "is a resource that can make silent images sharpen attention to the most subtle as to the most extreme forms of feeling. While it invokes and alludes to the poetic elegy, pictorial elegy also reaches into the silence before speech. A scream waiting to happen, a poem not yet written" (680). This tragic, silent scream is the melancholic cry of modern elegy. The turn to abstraction and nonobjective visual elegiac forms (exemplified in works such as Picasso's *Guernica* and Robert Motherwell's *Elegy to the Spanish Republic* series) recognizes that the mourned other who initially appeared lost – the uncountable and untraceable war dead, the Spanish Republic, and more recently, wilderness, bodies of water, entire species – becomes a lacunal absence to be perpetually mourned. No image of these others' once-living forms can efface the brutality of their destruction, nor should it. The responsibility of the living, as Woolf and her colleagues realized, is to refuse to allow the injustice and atrocities of past wars to be repeated in the future. In performing what Tammy Clewell (2004) calls an "anticonsolatory practice of mourning," whose "commemorative forms [are] intended to provoke and hurt, rather than console and heal," resistant elegies "[compel] us to refuse consolation, sustain grief, and accept responsibility for the difficult task of remembering the catastrophic losses of the twentieth century" (199). I believe that these aesthetic protests against forgetful complacency are part of an ethics of mourning that has a strong resonance and ongoing relevance in our fraught contemporary world. Politically and ethically engaged melancholic-elegiac visual works are fitting responses to the tragedies of the twentieth century. They refuse to be consoled, to let the pain be numbed by time; they refuse to obfuscate or forget. Would that humanity had heeded their voiceless cries.

Melancholia may be confused with depression or self-pity, and as such may be critiqued for creating paralysis rather than action, but I agree with Morton and Rae, who suggest that resistant elegy's melancholic tone can spur an ethical impulse to act. Morton (2010) stresses the importance of "a politicized melancholy" (255) and, as I mentioned above, Rae (2007) advocates "activist melancholia" (19) whose manifestation in resistant mourning practices "may be the basis for progressive political reform" (20). This purposeful melancholia echoes Judith Butler's (2004) "tarrying with grief ... remaining exposed to its unbearability," and provides an answer to her questions about what we might gain from maintaining an ongoing relationship with grieving: "are we ... returned to a sense of human vulnerability, to our collective responsibility for the physical lives of one another?" (30). There is indeed a great deal to be gained from a commitment to grieving, in particular when the losses are unjust and put at risk our collective future. I would add that, in my view, "one another" includes more-than-human others as integral participants/recipients of this collective responsibility. A confrontation with another's death (in the case of Robert Motherwell's *Elegies to the Spanish Republic* the death not of a person but a state, and by extension the death of justice and democracy; in the case of ecological elegy, the past and potential deaths of parts or all of the biosphere) reminds us of our own mortality, whose "acknowledgement names the condition for our ethical orientation in the world, the very condition, as Derrida puts it, of 'hospitality, love or friendship'" (Clewell 2004, 207). Derrida (1986) has written passionately about the ethical value in rejecting normative Freudian mourning and instead embracing the endless grieving that, in his view, comes with a full comprehension of the reality of a loved one's death.

Like Derrida and Butler, Clifton Spargo (2004) sees an ethical dimension in mourning, and particularly in inconsolable or melancholic mourning. When anti-elegiac artists or other mourners[2] refuse consolation or a redemptive vision, this dwelling on the death of the other may imply a responsibility to act – a "wishful intervention" (5) – for that other. Rae explains that this kind of ethical mourning "involves an acute and stubborn retroactive sense of responsibility for the loss" (18). According to Rae, contemporary AIDS activists such as Douglas Crimp and Michael Moon "have characterized any imperative to complete the 'work of mourning' for victims of AIDS as a call to endorse the *status quo*: an imperative to work for the restoration of 'normalcy,' in more ways than one. They have championed chronic melancholia on the grounds that it keeps things unsettled; it prevents a preventable catastrophe from becoming assimilated into the order of things" (18). Indeed, Spargo (2004) states that "an opposition to death … may function as the predicate of justice" (20). To be clear, this melancholic focus on death is not meant to be morbidly paralyzing – it is meant to energize and galvanize the survivors; in a sense, it is actually hopeful, driven by the vision of a future in which these wrongs have been righted. In its drive toward social and ecological justice, the deliberate and ethically motivated melancholia I am advocating moves us toward collectivity and action. This ethical imperative links modernist responses to the horrors of war to contemporary political and aesthetic forms of engagement with social injustice and looming environmental catastrophe (not yet a foregone conclusion, but an augury).

Ecological elegies mourn past losses such as habitat destruction and species extinction, and they also warn against the kind of absences we will be mourning in the future should present losses be allowed to continue. Here we may return to Morton's (2010) ecological "elegies for an incomplete process, elegies about events that have not yet (fully) happened … They fuse elegy and prophecy, becoming elegies for the future … Traditionally, elegies weep for that which has already passed. Ecological elegy weeps for that which *will have passed* given a continuation of the current state of affairs" (254). Rae uses the term "proleptic elegy"[3] to refer to elegies that were produced by artists between the world wars, when a sense of the looming possibility of another conflagration drove poets and other artists to create works of anticipatory grieving. She suggests that as a version of "ethical mourning" (such as Spargo discusses) "proleptic elegy could justly be added to the arsenal of resistant modes of mourning compiled in recent years by activists looking for social hope in devastating loss" (229). It can also be a mode of ethico-aesthetic praxis by artists engaged in ecological issues. While some of the artworks I will discuss engage in backward-gazing grieving and melancholia (works dealing with extinct species, for example), others, including my installation *Augury : Elegy*, are proleptic ecological elegies, grieving and warning about the kinds of future losses that could occur given a continuation of the status quo. These works are often visually arresting, even disturbing. In an era of over-saturation and media numbness, this often seems like the only way to effectively communicate the urgency of the situation. Indeed, Morton suggests that "progressive ecological elegy must mobilise some kind of choke or shudder in the reader that causes the environmental loss to stick in her throat, undigested. Environmental elegy must hang out in melancholia and refuse to work through mourning to the (illusory) other side" (256).

In what I see as an inheritance of the ethical melancholia of modernist elegies, the visual ecological elegies that I create and research attempt to stir viewers to reconsider their capacity to be what Mick Smith (2013) would call "ecologists," to become ethical participants in "ecological community" and prevent catastrophic loss on a biospheric scale:

> Perhaps, one might even say, the realisation of ecological community only begins to make sense through the senseless event of extinction. This pointless and irredeemable loss *touches* some of us in ways that reveal the infinite complications in trying to specify what is left in the wake of the death (finitude) of an entire mode of being(s). The ecologist (in a more than scientific sense) is someone who is touched by this loss in such a way as to mourn the toll of extinction instituted by human exemptionalism and exceptionalism. She is bereft and yet also understands that this feeling, her being touched by irrevocable loss, is itself a matter of realising the existence of a sense of an ecological *and* ethical *and* political community with other species. The species lost is not just a potential resource of which humans are deprived, but an example of *exceptional ethical irresponsibility*, one which can also incite (ethical) responsibilities and (political) resistance. (31)

Indeed, mourning, as James Stanescu (2012) writes in "Species Trouble: Judith Butler, Mourning, and the Precarious Lives of Animals," "is all about ethical, political, and ontological connections ... Mourning is a way of making connections, of establishing kinship, and of recognizing the vulnerability and finitude of the other" (568). This recognition implies responsibility. Butler (2003) herself claims that ongoing mourning for "the irrecoverable" – for what is past but in an ongoing present that also implicates the future – "becomes, paradoxically, the condition of a new political agency" (467). For ecological activists and artists, the articulation and enacting of this agency, of this politicization and ethical engagement rooted in mourning, can take many forms, from the social to the aesthetic, all of which demonstrate a sense of connection and community among and across species.

A 2014 study by the World Wildlife Fund and the Zoological Society of London found that the past forty years has seen a 50 per cent reduction in the number of wild vertebrate animals on Earth: "Creatures across land, rivers and the seas are being decimated as humans kill them for food in unsustainable numbers, while polluting or destroying their habitats" (Carrington 2014). And according to the Center for Biological Diversity (2015), current extinction rates are estimated at "1,000 to 10,000 times the background rate [of between one and five species annually], with literally dozens going extinct every day."[4] The report goes on to confirm the fact that, unlike past extinction events, this one is definitely anthropogenic: "99 percent of currently threatened species are at risk from human activities, primarily those driving habitat loss, introduction of exotic species, and global warming. Because the rate of change in our biosphere is increasing, and because every species' extinction potentially leads to the extinction of others bound to that species in a complex ecological web, numbers of extinctions are likely to snowball in the coming decades as ecosystems unravel." In an article in *The Guardian*, Jo Confino (2014) responds to such data, stating: "The question we should all be asking is why aren't we on the floor doubled up in pain at our capacity for industrial scale genocide of the world's species." In answer, he proposes grieving as "a pathway out of a destructive economic system." There is ample evidence that total biospheric calamity is a very real possibility – yet potentially, as Confino suggests, a preventable one. So to return to modern elegies: the ethical demands of this genre seem more potent and possible (and therefore hopeful) in the case of proleptic ecological elegy than in the case of past world wars and genocides – because there may still be time for us, even as ordinary individuals, to actively prevent the future deaths we may feel compelled to elegize and preemptively mourn. Oppositional, melancholic, political ecological mourning can inform environmental ethics and interventions, transforming the overwhelmed fatalistic impotence of "climate despair" into a more active, hopeful, and productive engagement with ecological issues – a refusal to acquiesce to the death of the ecological world and a commitment to justice for more-than-human as well as human communities and environments.

1844
Labrador Duck
Sea Mink
Arizona Jaguar
1905
Passenger Pigeon
1914
New Zealand Grayling
1923
Woodland Caribou
1935
Bali Tiger
1937
Cape Verde Giant Skink
1940
Texas Red Wolf

Figure 8.2 (opposite) | Gwen Curry, *Song of the Dodo* (1999).

Figure 8.3 (above) | Gwen Curry, *Song of the Dodo* (detail) (1999).

## Auguries

How might a genuinely radical "environmental ethics" be characterized? Perhaps as an attempt to express our *feelings* for the natural world in a way that speaks for that world's conservation.

SMITH 2005, 147[5]

If we cannot learn to listen, perceive and feel our ecological connection, to enlarge our boundaries of compassion, then we will continue to wreak havoc not only on birds but also on all living species, with dire consequences for the Earth. And, ultimately, for ourselves.

ANGELUCCI 2014

In an article entitled "What the Warming World Needs Now Is Art, Sweet Art," 350.org founder Bill McKibben (2005) writes, "if the scientists are right, we're living through the biggest thing that's happened since human civilization emerged. One species, ours, has by itself in the course of a couple of generations managed to powerfully raise the temperature of an entire planet, to knock its most basic systems out of kilter." But, he writes, this life-threatening catastrophe has not led to the kind of urgent, global-scale corrective and preventative action one might hope for. Scientists are able to measure the effects of climate change in various ways, but McKibben asks "can we register it in our imaginations, the most sensitive of all our devices?" This is where "art, sweet art" steps in: "We are all actors in this drama, more of us at every moment ... It may well be that because no one stands outside the scene, no one has the distance to make art from it. But we've got to try. Art, like religion, is one of the ways we digest what is happening to us, make the sense out of it that proceeds to action." Wendy Lynne Lee (2006) similarly argues that "justice-driven emancipatory action" that "foster[s] respect for biodiversity and ecological stability" can have its roots in both ecological and aesthetic realms, resulting in what she terms "a critical feminist political praxis capable of appreciating not only the value of human life, but those relationships upon which human and nonhuman life depend" (21). While direct-action activism is necessary, ethico-aesthetic-affective interventions are also required, in and out of galleries and institutions, created collaboratively with/in communities and also by individuals. Such interventions provide an answer to Rob Nixon's call in *Slow Violence and the Environmentalism of the Poor* to "convert into image and narrative the disasters that are slow moving and long in the making, disasters that are anonymous and that star nobody, disasters that are attritional and of indifferent interest to the sensation driven technologies of our image-world," and they begin to answer his question of how we might "turn the long emergencies of slow violence into stories dramatic enough to rouse public sentiment and warrant political intervention, these emergencies whose repercussions have given rise to some of the most critical challenges of our time" (2011, 3).

In 1964, Marshall McLuhan (1970) wrote "I think of art, at its most significant, as a DEW line, a Distant Early Warning system that can always be relied on to tell the old culture what is beginning to happen to it" (59). And more than simply telling, artworks can engage viewers or participants in profound ways. Ecologically engaged art is McLuhan's warning system in action, on a mission to help us redirect our self-destructive trajectory. When attempts at rationally explaining the urgency of climate issues and their relationship to our personal and societal choices and patterns of consumption have failed, the arts, as McKibben suggests above, may yet be able to effect change by engaging people's hearts and senses – tapping into empathy and feeling rather than intellect.[6] In her writing on the role of the artist in the age of ecological crisis, Beth Carruthers (2003) states that it must be one of "translator and messenger ... Art and art practices facilitate a process of learning through the engaged senses, one that can bypass conditioned patterns of thinking, allowing other ways of knowing to come forward, at times subtly, at times overwhelmingly." In *The Three Ecologies*, Félix Guattari (2000) proposes that we clarify the profound ecological questions of our time with a transversal "ecosophy" (19) whose methodologies are more artistic than scientific; in order to attain this ecosophical subjectivity, Guattari states, we must "forge new paradigms that are ... ethico-aesthetic in inspiration" (25). Suzi Gablik (2003) also sees possibilities for transformation in art practice, advocating transdisciplinary "eco-ventionist" artworks that situate "community at the core of our species nature" and foreground "empathic attunement." I see the resistant melancholic-elegiac sensibility and its urgent proleptic, anticonsolatory ethic as an aesthetic method for achieving this empathetic ecosophical attentiveness – one that is increasingly relevant in this era of ecological crisis brought on by "the old culture."

Figure 8.4 | Edward Burtynsky, *Oxford Tire Pile #2, Westley, California* (1999).

The eco-melancholic zeitgeist that the editors of this anthology have identified includes artworks that, in foregrounding melancholia for ecological losses (like the ethical anti-war grieving of the modernist elegists), attempt to open spaces for affective, empathetic, and ultimately ethical responses. They offer the possibility of transcending what Smith (2007) calls "human exemptionalism and exceptionalism" (or anthropocentrism), encouraging a sense of biospheric community and resultant positive ecological action on behalf of all (human and more-than-human) vulnerable and grievable beings.[7] This is part of a movement to work to heal widespread cultural ignorance of and disdain for the more-than-human, which has been subject to what Rae terms "systematic occlusions" (20) from the sphere of the grievable. What follows is a discussion of a small selection of such ecologically engaged artists and artworks.

Figure 8.5 | Christian Boltanski, *Personnes* (2010), installation view, Grand Palais, Paris.

Edward Burtynsky is internationally known for his large photographs of the effects of industrialization on the earth. The images in his *Oil* series in particular could be read as elegies that resonate with the "quantitative sublime" (to borrow from Kant) that can also be seen in some anti-elegiac modernist and contemporary artworks such as *Paths of Glory* by C.R.W. Nevinson (1917), whose ironic title was taken from Thomas Gray's poem "Elegy Written in a Country Church-Yard" (1750), and *Personnes* by Christian Boltanski (2010); these works confront the viewer with overwhelming numbers of dead objects – or people – in an eternally present melancholic moment. Jonathan W. Marshall (2011) also reads mourning and melancholia into Burtynsky's work: "Oscillating between formalistic containment and abject disintegration, Burtynsky's oeuvre performs an act of mourning for the Modernist subject and the visual and industrial tropes which once sustained it, producing a

Figure 8.6 | Gwen Curry, *Void Field (after Kapoor)* (1998–2002).

series of theatrical relations which are at once melancholy and critically engaged, provocative and nostalgic, disturbing and formalistically reassuring." Burtynsky's painterly photographs illustrate with astonishing clarity the near-unimaginable scale and scope of industrialization, and in images such as *Oxford Tire Pile #2*, mounds of used tires evoke the unburied dead that haunt modernist anti-war elegies as a warning to those who remain alive.

Echoing this tendency toward quantification, Gwen Curry's *Song of the Dodo* (1999) is an explicitly mournful ecological memorial, sombrely bearing the names of extinct species and their extinction dates. Curry (2009) writes that "There is no room for irony in this very dark work." She quotes Derek Besant's commentary in the catalogue for her 2001 solo exhibition, *Witness*: "There are no bones, no fingerprints, no photographs, or tracks in plaster. We must attempt to recall that which

Figure 8.7 | Gwen Curry, *Void Field (after Kapoor)* (detail) (1998–2002).

has already been removed. We are left to look for the missing pieces within ourselves." Curry's installation of memorial tiles, *Void Field (after Kapoor)*, performs a similar work of mourning: "*Void Field* creates a sacred place of remembrance much like the floors of cathedrals whose crypts honour the deceased. Each of the tiles bears the name and date of extinction of a bird, mammal, fish, reptile or amphibian. This project was sad and humbling for me to do and it became a kind of atonement for our collective loss. When the work was shown in galleries there was a respectful hush and on one occasion when I was present a viewer burst into tears" (ibid.). Curry's memorial works extend grievability to these forever-lost others, bringing the finality of their species death to the present. There is also a tacit sense of culpability that sensitive viewers may feel, knowing that most or all of those species (especially fabled ones such as the dodo and passenger pigeon) perished as a direct result of human activity and asymmetrical relations with the more-than-human world, and knowing that this imbalance persists today. As such, this elegiac work adds to the narratives of loss that Nixon calls for in *Slow Violence* – the kinds of stories that Thom van Dooren (2014) claims can communicate ecological loss in such a way as to create a feeling of "affective involvement in a loss"; citing Derrida, van Dooren asserts that this kind of "storied-mourning" does not seek consolatory strategies for recovering from loss, but rather "offers us the possibility of mourning as a deliberate act of *sustained* remembrance that requires us to interrogate how it is that we might 'live *with* ghosts' … This is the kind of mourning that asks us – that perhaps demands of us, individually and collectively – to face up to the dead and to our role in the coming into being of a world of escalating suffering, loss, and extinction" (142–3).

Addressing current losses rather than past extinctions, and originating in my own ecological grieving, my installation, entitled *Augury : Elegy* (2011), viscerally and proleptically warns about species loss, with hundreds of bones standing in for the hundreds of millions of birds dying annually due to human activity. The following text accompanied the installation:

> Just before midnight on New Year's Eve 2011, in Beebe, Arkansas, 4,000 or so blackbirds fell out of the sky, dead. Around the same time, several hundred grackles, redwing blackbirds, robins, and starlings dropped dead in Murray, Kentucky. A few days later, 500 dead blackbirds, brown-headed cowbirds, grackles, and starlings were found on a highway in Pointe Coupee, Louisiana, while 200 dead American coots appeared on a bridge in Big Cypress Creek, Texas. On January 4, in Falköping, Sweden, 100 jackdaws were found dead in the street. And then on January 5, some 1,000 dead turtle doves rained down on the town of Faenza in Italy. Later that year, on October 23, 6,000 dead birds washed up on the southeastern shore of Ontario's Georgian Bay, and then, remarkably, Beebe was again showered with the bodies of 5,000 blackbirds on New Year's Eve 2012.
>
> It seems a little apocalyptic.

> One might well ask whether this series of mass deaths is a microcosm of humanity's increasingly toxic impact on the non-human world. But we are not just poisoning an isolated wilderness "out there." We are poisoning *our* ecosystems – our sources of food, water, and air; our only home. The warnings are everywhere, if only we would choose to see and heed them. Because those were a lot of canaries, and we're all in this coalmine together.

Augury refers to signs and omens, and relates to the ancient Roman tradition of interpreting the behaviour (or the bones or guts) of birds as a sign of divine approval or disapproval. This piece warns that the current ecological crisis will augur an elegy if humanity continues on the destructive path of modernity. According to Erikson et al. (2001), anthropogenic avian mortality (resulting from human activities excluding food production) "may total between 100 million and 1 billion birds per year in the United States alone" (in Podolsky 2010, 11). In addition to that staggering number bird populations worldwide are being negatively impacted by climate change,[8] which by altering temperatures, ecosystems, and habitats affects birds' lives and population sizes (Miller-Rushing 2010). Alongside species (and species groups or genera) such as bees, bats, corals, and fish, birds are "crucial indicators of ecosystem health. Changes in bird populations signal changes in the ecosystems we depend on for vital environmental services such as food, clean air, and water" (BirdLife International 2012). The storm of bones (salvaged from discarded chicken carcasses) in my project also hints at the ecological problem of the over-consumption of meat, in particular that produced in factory farms, which, in addition to issues of animal ethics, are major polluters.[9] In this work, I am trying to communicate the kind of resistant ecological mourning that motivates awareness and sustainable practices.[10] Because we must not simply remember, we must act. While *Augury : Elegy* has a strongly melancholic, elegiac tone, it is my ultimate hope – to reiterate Rae's point – that this ethic of melancholic mourning carries with it possibilities for positive social change.

In the vein of avian mortality and bones, Deborah Samuel's photo series entitled, fittingly, *Elegy* (2012) is a deathly portrait gallery of animal skeletons and various bones, eerily illuminated against pitch-black

Figure 8.8 | Jessica Marion Barr, *Augury : Elegy* (2011).

Figure 8.9 | Jessica Marion Barr, *Augury : Elegy* (detail) (2011).

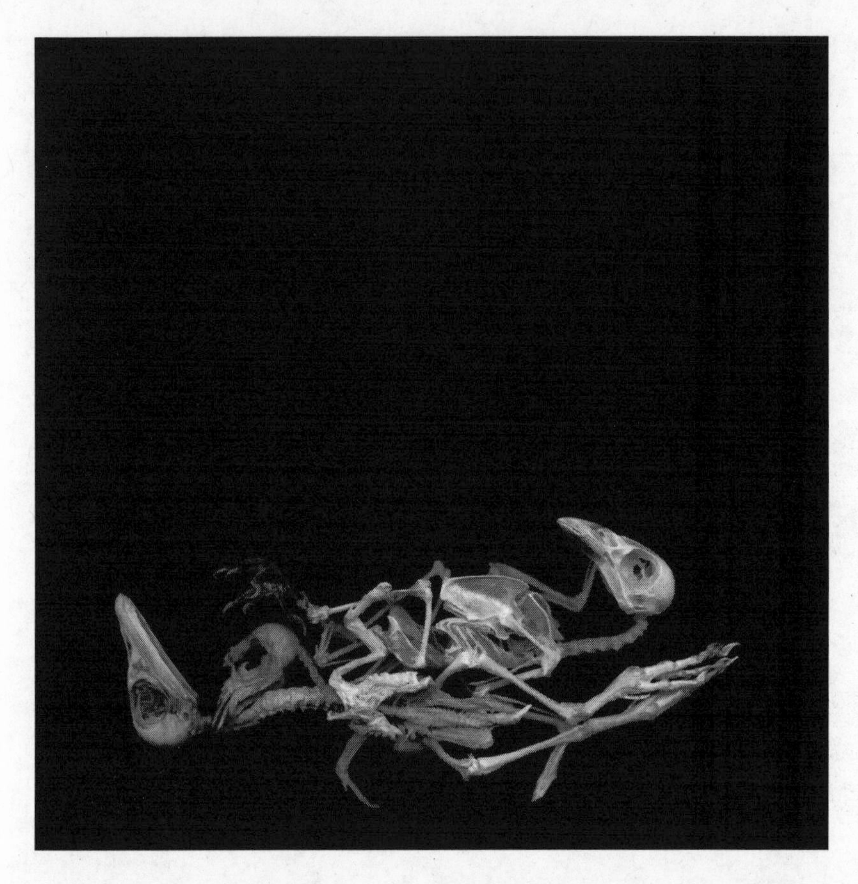

Figure 8.10 | Deborah Samuel, *Elegy – Cardinal, Duck and Solitaire.1*, from the *Elegy* series, produced with the Royal Ontario Museum (2012).

backgrounds. This series was produced in conjunction with the Royal Ontario Museum in Toronto, and the remains were photographed using scanners (hence the otherworldly glow that the bones seem to emanate). All of the photographs in the exhibit were accompanied by didactic panels identifying the species and, in some cases, commenting on the ecological threats that they are facing. Samuel's inspiration for the series began with her pained response to the images of oil-coated pelicans in the wake of the Gulf of Mexico oil spill in 2010. According to Sarah Hampson, Samuel "wanted to photograph the stricken birds to stir public consciousness about society's failure to protect other living things. But the government prohibited access to the site. And then she

Figure 8.11 | Chris Jordan, CF000144 from the series *Midway: Message from the Gyre* (2009–present).

thought of another way to use birds to suggest the life they once embodied" (2012). Matthew Wright's (2012) interview with Samuel reveals that the artist "not only tried to imagine the relationships these animals had in life, but more importantly, those they would've had if not victimized by manmade disaster." In reanimating these bones, Samuel refuses to bury the dead, to be resigned to unjust ecological losses; her elegies subtly resist these losses, and encourage us to do the same.

Chris Jordan's *Midway: Message from the Gyre* photo series takes this macabre gaze to an even more visceral documentary level. He writes:

> On Midway Atoll, a remote cluster of islands more than 2000 miles from the nearest continent, the detritus of our mass consumption surfaces in an astonishing place: inside the stomachs of thousands of dead baby albatrosses. The nesting chicks are

Figure 8.12 | Brandon Ballengée in scientific collaboration with Stanley K. Sessions, with versified titles forming a poem by KuyDelair, *Vertical Fall in the Winter call that dances in the spring nocturnal ...* (2010/2012), from the series *A Season in Hell, Deadly Born Cry* (unique digital chromogenic print; cleared and stained hatchling song sparrows found dead of an unknown cause in the wild).

fed lethal quantities of plastic by their parents, who mistake the floating trash for food as they forage over the vast polluted Pacific Ocean. For me, kneeling over their carcasses is like looking into a macabre mirror. These birds reflect back an appallingly emblematic result of the collective trance of our consumerism and runaway industrial growth. Like the albatross, we first-world humans find ourselves lacking the ability to discern anymore what is nourishing from what is toxic to our lives and our spirits. Choked to death on our waste, the mythical albatross calls upon us to recognize that our greatest challenge lies not out there, but *in here*. (2011, emphasis in original)

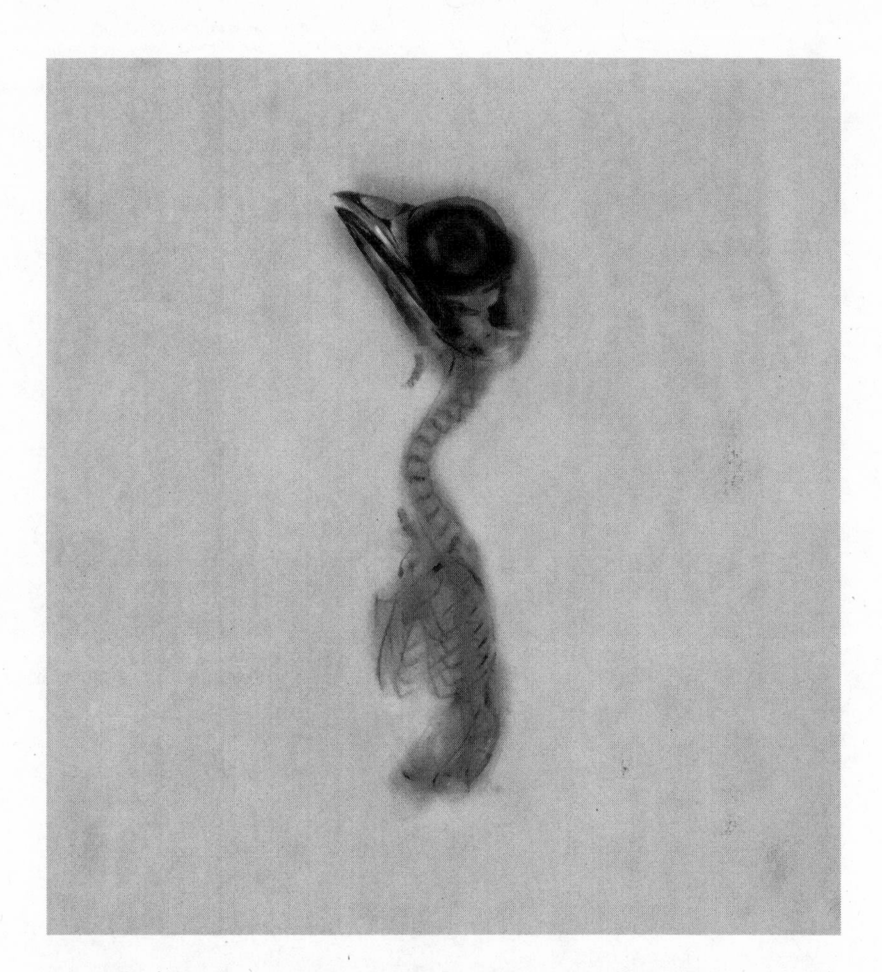

Figure 8.13 | Brandon Ballengée in scientific collaboration with Stanley K. Sessions, with versified titles forming a poem by KuyDelair, *Moonlight chant in the metamorphosis hanged up to the Air* (2010/2012), from the series *A Season in Hell, Deadly Born Cry* (unique digital chromogenic print; cleared and stained laboratory-reared research domestic chicken embryo with missing wings and legs).

These images of dead baby birds, revealing bellies full of recognizable plastic items, elicit strong reactions – shock, disgust, sadness, grief, regret – in most viewers. Held in an eternal present, their corpses are not given the opportunity to participate in the consolatory cycle of decay and rebirth through the composting processes of "nature." Jordan's images are truly melancholic, resistant elegies that refuse to bury this ongoing tragedy.

Figure 8.14 | Brandon Ballengée with Todd Gardner, Jack Rudloe, Brian Schiering, and Peter Warny, *Collapse* (detail) (2012), mixed-media installation including 26,162 preserved specimens representing 370 species from the Gulf of Mexico, South Gallery Installation shot.

Like Jordan, biologist and artist Brandon Ballengée exhibits disturbing, morbid, and melancholic images (and even specimens) that clearly demonstrate the toxic impact of industrialized humanity on the more-than-human world. His recent exhibition, entitled *Collapse: The Cry of Silent Forms*, brings his research and art together, transmuting Ballengée's "field research into metaphors that reveal the fragility of life forms in degraded ecosystems" (Ronald Feldman Fine Arts 2012). *Collapse* is a large installation consisting of 26,162 preserved aquatic specimens in glass jars stacked to form a pyramid. These specimens represent 370 species from the Gulf of Mexico, where Ballengée and his colleagues have been conducting research on "the global crisis of the world's fisheries and the current threat for the unraveling of the Gulf of Mexico's food-chain following the BP oil spill. The large-scale installation ... recalls the fragile inter-relationships between Gulf species. Empty containers represent species in decline or those already lost to extinction" (Ronald Feldman Fine Arts 2012). His other works, *A Season in Hell Series: Deadly Born Cry* (2010/2011), and *Malamp: The Occurrence of Deformities in Amphibians* (1996–current), include scanned photographs of dead and deformed birds and frogs, victims of ecological contamination and laboratory experimentation. The title of one of the amphibian pieces, *Un Requiem pour Flocons de Neige Blessés (A Requiem for Wounded Snowflakes)* (2009/2011), reinforces the elegiac tone of these haunting, disturbing, deathly images. Like the artists I have mentioned above, Ballengée vividly presents ecological imagery that invites viewers to pause, reflect, and grieve.

There are many responses to modernity, industrialization, climate change, and ecological catastrophe (which is inextricably linked to human catastrophe). Grieving is one very real response. In expressing perpetual (albeit resistant and hopefully politically potent) mourning, the artworks I have described – like the resistant elegies of modernist artists coping with the damage done by the World Wars – give voice to feelings that many of us have but do not know how to express, the emotional dimensions of an unimaginably vast loss. This can be part of a process of healing, starting with the first link in a chain described in the NAMES Project AIDS Memorial Quilt's motto: "Remember Understand Educate Act" (Quilt2012, 2012).[11] While projects like Jordan's and Ballengée's also participate in aiding understanding through pedagogical components, elegiac works focused on the uncomfortable act of remembering and representing ecological loss can aid in a potentially cathartic articulation of individual or collective pain, an act of feeling which can be the essential first step in a journey toward more positive and proactive ways of addressing the great problems of our time. In order to be restored to an ethical, ecological, and ultimately hopeful sense of responsibility,

Figure 8.15 | The NAMES Project, AIDS Memorial Quilt, Washington, 1992 (1987–present).

we must soberly confront the past and present; we must remember the damage and our various roles in it – and this is what the ethico-aesthetic work of ecological elegy aims to do. "Now is a time for grief to persist, to ring throughout the world," says Morton (2007, 185), echoing Joanna Macy and other ecological activists in a call to embrace the elegiac melancholia of proleptic, resistant ecological mourning – to insist on *remembering* as a "crucial task of continuous mindfulness" (Macy 2012, 223) in order to prevent further damage and to create a better world for future life on earth. The artworks I have discussed here elegize past, present, and potential future losses, holding viewers in a present-continuous of ecological death and anthropogenic loss – inviting us to feel a persistent and resistant grief. I believe that the empathetic space opened by ecological elegy can remind viewers and participants that we are all part of earth's ecological community, with a responsibility to care for and preserve all of the beings who share this biosphere. Or, less ambitiously, I would suggest that even brief moments of engagement and feeling about and for a more-than-human ecological community are a vital component of confronting the reality of this stage of the Anthropocene.

Mick Smith (2005) stresses the value of affect and "structures of feeling" in his discussions of environmental ethics, and ecologically elegiac artists work primarily from a place of affect that is infused with factual and theoretical information from our research. Nicole Shukin (2012) has emphasized the role of artists in "mak[ing] amorphous calamities visible" and Norah Bowman-Broz (2012) states that "artists draw us in to the affective intensity of these intense [ecological] events." As maps of the affective terrain of our fraught relationships to the ecological world in crisis, these projects create space for an expression of how many of us feel about climate change, responsibility, ecological loss, and the future of the biosphere. Morton asks: "Isn't this lingering with something painful, disgusting, grief-striking, exactly what we need right now, ecologically speaking?" (197). This work is indeed painful and mucky, inhabiting a place of acute awareness, melancholic mourning, and ethical grief. Fredric Jameson (1991) states that the role of the political artist is to create place, to help us locate ourselves against the disorientation and confusion of late capitalism – and this is part of the contribution that melancholic ecological elegies make. If we can locate ourselves in a place of affective kinship with the more-than-human world, we will be in a position to be interpellated or pulled, as Levinas would have it, by the ecological Other, expanding our ethical orientation to include care for those Others with whom we may have nothing in common.

This opening to the ecological other is the crux of ecological artwork; it is the intersection of ethics and affect, which is the territory of resistant mourning and proleptic elegy. In the summer of 2012, I gave a conference presentation which included a performance of sorts: with images of my artwork projected in a darkened room, I walked around carrying a cloth bag containing chicken bones, while reading my text about the spate of mass bird-flock deaths in 2011–12. I invited each participant to take a bone, to examine it, and to reflect on the feelings that arose in them in relation to their experience with the bone while listening to the monologue and viewing an image of *Augury : Elegy*. Afterwards, a visibly

Figure 8.16 | Karen Abel and Jessica Marion Barr, *Vernal Pool: A Participatory Art Project About Place + Precipitation* (installation detail) (2014). Approximately 2,000 pieces of salvaged glassware filled with samples of melted snow collected by 114 participants from six countries; see http://vernal-pool.tumblr.com.

moved participant commented that he had experienced a new and deep empathy for "these creatures whom I've eaten hundreds of times without ever really thinking about them." He was moved to reconsider his relationship with these ecological others, to extend grievability to them, and more broadly to see this as an entry point into action on behalf of the future of our biosphere. This moment offers a glimpse at the possibility of ecological artworks creating space for the feeling that more-than-human others *matter*, which according to Freya Mathews (2011) is at the heart of environmental ethics.

So, to end on a less melancholic note, I believe that we can change the picture, but only if we recognize that there is a problem. Ecologically elegiac artworks attempt this intervention, providing a new and often horrible view – one that might make us *feel* in such a way that we cannot help but act to prevent the apocalyptic visions we are seeing. But we also need a vision of what we want – a positive goal to work for, not just a negative image to fight against. This is where collaborative, hopeful eco-art steps in – bioremedial, utopian, community-based, participatory – creating new possibilities for a future of symbiotic collaboration and cooperation within this ecological community.

## NOTES

1 See, for example, Pablo Picasso's *Guernica* (1937), Robert Motherwell's *Elegies to the Spanish Republic* (c. 1949–85), and Anselm Kiefer's massive mournful German landscapes (1970s–80s).

2 For example Hamlet. Rae also mentions the "militant sadness" of Argentinian and Serbian mothers (2007, 18). Artist Marina Abramovic's video installation and performance piece *Balkan Baroque* (1997) is another compelling and disturbing example of melancholic mourning (through artwork), in this case for the war in Bosnia, but according to the artist by extension for "any war, anywhere in the world" (www.moma.org/explore/multimedia/audios/190/1988). In the piece, Abramovic sat atop a pile of 1,500 bloody cow bones, scrubbing each one clean. The pile of cleaned bones was then left in the gallery in front of a wall where a three-channel video dealing with Abramovic's experience of the Bosnian war and her Yugoslavian heritage played.

3 "Prolepsis" essentially means "anticipation." It can also refer to "the representation of a thing as existing before it actually does or did so" – e.g. "he was a dead man when he entered" (Oxford American Dictionary 2013). See Rae's article "Double Sorrow: Proleptic Elegy and the End of Arcadianism in 1930s Britain," *Modernism and Mourning*, 213–34.

4 Joanna Macy and Chris Johnstone (2012) write that "a third of all amphibians, at least a fifth of all mammals, and an eighth of all bird species are now threatened with extinction" (23).

5 Some italics removed.

6 This point has been made in many contexts and by many writers. My source here is Theresa May's keynote presentation "'This Is My Neighborhood!' – Community Identity, Ecology and Performance" at *Staging Sustainability*, a conference at York University, 20 April 2011.

7 The notion of grievability is most notably presented in Judith Butler's *Frames of War: When Is Life Grievable?*

8 See, for example, Andrew E. McKechnie and Blair O. Wolf, 2010, "Climate Change Increases the Likelihood of Catastrophic Avian Mortality Events during Extreme Heat Waves."

9 Over-consumption of factory-farmed meats is a serious environmental problem. See, for example, "Livestock a Major Threat to Environment," FAO *Newsroom*, 29 November 2006. http://www.fao.org/newsroom/en/news/2006/1000448/index.html. Accessed 12 April 2011.

10 As such, I aim to use sustainable materials in my art practice. *Augury : Elegy* is made from branches from my backyard, crochet thread inherited from a great-great-aunt, and chicken bones that were being discarded by a local grocery store. In order to clean the bones, I boiled them and made about forty-five bowls of chicken soup, which I gave to friends and neighbours.

11 The Quilt is a melancholic project that insists on remembering, even dwelling on, death as an ethical and activist gesture. A highly successful awareness- and fund-raising endeavour, the quilt has over 48,000 panels featuring over 94,000 names, and has been viewed by over 18 million visitors (NAMES Project Foundation 2011).

## REFERENCES


Angelucci, Sara. 2014. "A Mourning Chorus." http://sara-angelucci.ca/filter/Projects/A-Mourning-Chorus2014. Accessed 30 March 2015.

Bennett, Jill. 2012. *Living in the Anthropocene*. Monograph for dOCUMENTA (13): *100 Notizen – 100 Gedanken* No. 053. Ostfildern: Hatje Cantz Verlag.

BirdLife International. 2012. "First-Ever 'State of Canada's Birds' Report Released." *BirdLife International*. Published 27 June 2012. http://www.birdlife.org/community/2012/06/first-ever-state-canadas-birds-report-released/. Accessed 13 July 2012.

Blake, William. 2005. "Auguries of Innocence." In *William Blake: Selected Poems*, edited by G.E. Bentley, Jr., 295–8. London: Penguin Classics.

Bowman-Broz, Norah. 2012. "Radical Re-alignments in Human Communities: Living with the Anomalous Mountain Pine Beetle in Northern B.C." Presentation at *Space + Memory = Place*, the biannual conference of the Association for Literature, Environment, and Culture in Canada (ALECC), UBC Okanagan, Kelowna, 11 August.

Butler, Judith. 2003. "Afterword: After Loss, What Then?" In *Loss*, edited by David L. Eng and David Kazanjian, 467–73. Berkeley: University of California Press.

– 2004. *Precarious Life: The Powers of Mourning and Violence*. London: Verso.

– 2010. *Frames of War: When Is Life Grievable?* New York: Verso.
Carrington, Damian. 2014. "Earth Has Lost Half of Its Wildlife in the Past 40 Years, Says WWF." *The Guardian*. Last modified 30 September 2014. http://www.theguardian.com/environment/2014/sep/29/earth-lost-50-wildlife-in-40-years-wwf. Accessed 25 October 2014.
Carruthers, Beth. 2003. "About Cultural Currency." *Cultural Currency*. http://www.culturalcurrency.ca/AboutCulturalCurrency.html. Accessed 4 December 2011.
Center for Biological Diversity. 2015. "The Extinction Crisis." http://www.biologicaldiversity.org/programs/biodiversity/elements_of_biodiversity/extinction_crisis/. Accessed 16 March 2015.
Clewell, Tammy. 2004. "Consolation Refused: Virginia Woolf, the Great War, and Modernist Mourning." *MFS: Modern Fiction Studies* 50, no. 1: 197–223.
Confino, Jo. 2014. "Grieving Could Offer a Pathway out of a Destructive Economic System." *The Guardian*. Last modified 2 October 2014. http://www.theguardian.com/sustainable-business/2014/oct/02/grieving-pathway-destructive-economic-system. Accessed 25 October 2014.
Curry, Gwen. 2009. "Major Works: *Song of the Dodo*." Last modified 2009. http://www.gwencurry.com/mw-song-dodo.html. Accessed 1 May 2013.
– 2009. "Major Works: *Void Field (after Kapoor)*." Last modified 2009. http://www.gwencurry.com/mw-void-field.html. Accessed 1 May 2013.
Derrida, Jacques. 1986. *Mémoires: For Paul de Man*. Translated by Cecile Lindsay, Jonathan Culler, and Eduardo Cadava. New York: Columbia University Press.
Ellenborgen, Josh. 2010. "On Photographic Elegy." In *The Oxford Handbook of the Elegy*, edited by Karen Weisman, 681–99. Oxford: Oxford University Press.
Food and Agriculture Organization of the United Nations. 2006. "Livestock a Major Threat to Environment: Remedies Urgently Needed." *FAO Newsroom*. Last modified 29 November. http://www.fao.org/newsroom/en/news/2006/1000448/index.html. Accessed 12 April 2011.
Freud, Sigmund. 1957. "Mourning and Melancholia." In *The Complete Psychological Works of Sigmund Freud, Vol. XIV (1914–1916)*, translated by James Strachey, 243–58. London: Hogarth Press.
Gablik, Suzi. 2003. "Alternative Aesthetics." *Landviews.org, Online Journal of Landscape, Art, & Design*. Last modified summer 2003, n.p. http://www.landviews.org/la2003/alternative-sg.html. Accessed 1 November 2012.
Gilbert, Sandra M. 2006. *Death's Door: Modern Dying and the Ways We Grieve*. New York: Norton.
Grierson, John. 1996. "First Principles of Documentary." In *Imagining Reality: The Faber Book of Documentary*, edited by K. MacDonald and M. Cousins, 101. London: Faber and Faber.

Guattari, Félix. 2000. *The Three Ecologies*. Translated by Ian Pindar and Paul Sutton. London: The Athlone Press.

Hampson, Sarah. 2012. "Why Deborah Samuel Photographed Animal Skeletons." *The Globe and Mail*, 21 May. http://www.theglobeandmail.com/life/relationships/why-deborah-samuel-photographed-animal-skeletons/article4198244/?cmpid=rss1. Accessed 1 May 2013.

Haraway, Donna. 2004. "Otherworldly Conversations; Terran Topics; Local Terms." In *The Haraway Reader*, 125–50. New York: Routledge, 2004.

Helsinger, Elizabeth. 2010. "Grieving Images: Elegy and the Visual Arts." In *The Oxford Handbook of the Elegy*, edited by Karen Weisman, 658–80. Oxford: Oxford University Press.

Homans, Peter, ed. 2000. *Symbolic Loss: The Ambiguity of Mourning and Memory at Century's End*. Charlottesville: University of Virginia Press.

Jameson, Frederic. 1991. "Postmodernism; or, The Cultural Logic of Late Capitalism," n.p. http://www.marxists.org/reference/subject/philosophy/works/us/jameson.htm. Accessed 1 November 2012.

Jordan, Chris. 2011. "Midway: Message from the Gyre." Last modified 2011. http://www.chrisjordan.com/gallery/midway/#about. Accessed 1 May 2013.

– 2013. "Midway." http://www.midwayfilm.com/ and http://www.midwayfilm.com/about.html. Accessed 20 February 2013.

Kagan, Sacha. 2011. *Art and Sustainability: Connecting Patterns for a Culture of Complexity*. Bielefeld: Transcript Verlag.

LaCapra, Dominick. 1999. "Trauma, Absence, Loss." *Critical Inquiry* 25: 696–727.

Lee, Wendy Lynne. 2006. "On Ecology and Aesthetic Experience: A Feminist Theory of Value and Praxis." *Ethics & the Environment* 11, no. 1: 21–41.

Levinas, Emmanuel. 1974. *Otherwise than Being, or, Beyond Essence*. Boston: Kluwer Academic Publishers.

– 1998. *Entre Nous*. Translated by Michael B. Smith and Barbara Harshav. London: Athlone.

Macy, Joanna, and Chris Johnstone. 2012. *Active Hope*. Novato, CA: New World Library.

Marshall, Jonathan W. 2011. "Corporeal Spectacle and Sublime Annihilation: Post World War Two US Art and the Theatrical Operations of Edward Burtynsky's Environmental Photography." *Double Dialogues: Boom or Bust: Economies of Production & Exchange in Theatre, Performance and Culture* 14 (Summer): n.p. http://www.doubledialogues.com/issue_fourteen/Marshall.html. Accessed 15 May 2013.

Martin, Russell. 2002. *Picasso's War: The Destruction of Guernica, and the Masterpiece That Changed the World*. New York: Dutton.

Mathews, Freya. 2011. "Planet Beehive." *Australian Humanities Review* 50 (May): 159–78.

McKay, Don. 2009. "Introduction: Great Flint Singing." In *Open Wide a Wilderness: Canadian Nature Poems*, edited by Nancy Holmes, 1–31. Waterloo: Wilfrid Laurier University Press.

McKechnie, Andrew E., and Blair O. Wolf. 2010. "Climate Change Increases the Likelihood of Catastrophic Avian Mortality Events during Extreme Heat Waves." *Biology Letters* 6: 253–6.

McKibbon, Bill. 2005. "Imagine That: What the Warming World Needs Is Art, Sweet Art." *Grist Magazine*, 21 April. http://www.grist.org/article/mckibben-imagine. Accessed 30 November 2012.

McLuhan, Marshall. 1964. *Understanding Media*. Quoted in Sam Black. 1970. "Herbert Read: His Contribution to Art Education and to Education through Art." In *Herbert Read: A Memorial Symposium*, edited by Robin Skelton, 57–65. London: Methuen & Co.

Miller-Rushing, Abraham J., Richard B. Primack, and Cagan H. Sekercioglu. 2010. "Conservation Consequences of Climate Change for Birds." In *Effects of Climate Change on Birds*, edited by Anders Pape Moller, Wolfgang Fiedler, and Peter Berthold, 295–6. Oxford: Oxford University Press.

Moller, Anders Pape, Wolfgang Fiedler, and Peter Berthold. 2010. *Effects of Climate Change on Birds*. Oxford: Oxford University Press.

Morton, Timothy. 2007. *Ecology without Nature: Rethinking Environmental Aesthetics*. Cambridge, MA: Harvard University Press.

– 2010. "The Dark Ecology of Elegy." In *The Oxford Handbook of the Elegy*, edited by Karen Weisman, 251–71. Oxford: Oxford University Press.

– 2011. "Unsustaining." *World Picture* 5 (Spring): 1–10. http://www.worldpicturejournal.com/WP_5/PDFs/Morton.pdf. Accessed 1 December 2012.

NAMES Project Foundation. 2011. "The AIDS Memorial Quilt." http://www.aidsquilt.org/about/the-aids-memorial-quilt. Accessed 12 October 2012.

Nixon, Rob. 2011. *Slow Violence and the Environmentalism of the Poor*. Cambridge, MA: Harvard University Press.

Podolsky, Richard. 2010. *2010 Initial Bird Assessment of the Dennis Water District Wind Turbine Project Site*. Arlington: Boreal Renewable Energy Development.

Quilt2012. 2012. "AIDS Memorial Quilt." http://quilt2012.org/. Accessed 12 October 2012.

Rae, Patricia. 2007. "Introduction: Modernist Mourning." In *Modernism and Mourning*, edited by Patricia Rae, 13–49. Lewisburg: Bucknell University Press.

– 2007. "Double Sorrow: Proleptic Elegy and the End of Arcadianism in 1930s Britain." In *Modernism and Mourning*, edited by Patricia Rae, 213–38. Lewisburg: Bucknell University Press.

– 2010. "'Between the Bullet and the Lie.'" In *The Oxford Handbook of the Elegy*, edited by Karen Weisman, 305–23. Oxford: Oxford University Press.

Ramazani, Jahan. 1994. *Poetry of Mourning: The Modern Elegy from Hardy to Heaney*. Chicago: University of Chicago Press.
Ronald Feldman Fine Arts. 2012. "Brandon Ballengée Collapse: The Cry of Silent Forms." http://www.feldmangallery.com/pages/exhsolo/exhbrandon12.html. Accessed 1 May 2012.
Shukin, Nicole. 2012. Plenary talk at *Space + Memory = Place*, the biannual conference of the Association for Literature, Environment, and Culture in Canada (ALECC), UBC Okanagan, Kelowna, 9 August 2012.
Smith, Mick. 2001. "Environmental Anamnesis: Walter Benjamin and the Ethics of Extinction." *Environmental Ethics* 23, no. 4: 359–76.
– 2005. "Citizens, Denizens and the Res Publica: Environmental Ethics, Structures of Feeling and Political Expression." *Environmental Values* 14, no. 2: 145–62.
– 2013. "Ecological Community, the Sense of the World, and Senseless Extinction." *Environmental Humanities* 2: 23–43.
Stanescu, James. 2012. "Species Trouble: Judith Butler, Mourning, and the Precarious Lives of Animals." *Hypatia* 27, no. 3: 567–82.
Van Dooren, Thom. 2014. *Flight Ways: Life and Loss at the Edge of Extinction*. New York: Columbia University Press.
World Wildlife Fund. 2012. "Know Your Flagship, Keystone, Priority and Indicator Species." http://wwf.panda.org/about_our_earth/species/flagship_keystone_indicator_definition/. Accessed 15 May 2012.
Wright, Matthew. 2012. "Deborah Samuel's Elegy Casts a New Light on Death." *National Post,* 2 April. http://arts.nationalpost.com/2012/04/04/deborah-samuels-elegy-casts-a-new-light-on-death/. Accessed 1 May 2013.

# 9 Making Loss the Centre

## *Podcasting Our Environmental Grief*

ANDREW MARK AND AMANDA DI BATTISTA

CATRIONA SANDILANDS: There is a particularization of loss ... and there is also a politicization of that loss ... what do we do with the loss? How do we make the fact of having lost part of a political response?

RALPH CARL WUSHKE: [When] I would get news that another person with HIV, with AIDS, had died, a friend of mine had died ... I would immediately sit down at my piano and play my favourite hymn, "For All the Saints," sing all eight stanzas, play it badly, and then write his name on that page in the hymnal. So this was my personal little ritual that I would immediately do when somebody died of AIDS. You know, that's sitting with the grief. This is a moment to cry, and a moment to sing, and there's nothing else to do here now. But it's a way of processing and moving through it. Certainly, by analogy, there is something like this to be done, whether it's an Earth quilt, or songs of lament for the species that are gone.

MINUTE 14:34,[1] "MAKING LOSS THE CENTRE" PART 2, *COHEARENCE*

## Introduction

As members of a large interdisciplinary environmental studies faculty at York University, we are continually confronted by the devastating effects of environmental loss and by the inability of our community of scholars, students, and activists to grieve for these losses in meaningful ways. In our teaching practices, we uncover systemic forms of violence against both humans and the more-than-human world and risk encouraging abject despair as we struggle to resist hegemonic Western capitalist narratives of progress. While the extinction of species across the globe, the violence of resource extraction, and the consequences of our daily consumptive practices are inscribed on the planet's surface, in our waters, and throughout the sky, they are also written onto our sensory being in complex ways. The unimaginable task of reckoning so many lives, places, and things lost often leaves environmentalists struggling to cope. Rather than wallowing in negativity, or insisting on opportunity, we – and others – feel that confronting and acknowledging loss can offer abundant, productive, social, and even militant modes of reimagining the environmental movement.

In 2012, we launched the podcast series *CoHearence*,[2] which we co-produce with the aim of exploring the intersections of culture and the environment. In our first episodes, "Melancholy, Mourning, and Environmental Thought: Making Loss the Centre" ("Making Loss the Centre," parts 1, 2), we interviewed six scholars – Peter Timmerman, Catriona Sandilands, Susan Moore, Ella Soper, Ralph Carl Wushke, and Honor Ford-Smith – working on very different areas related to environmental loss. We then wove their stories together to create a narrative of insights on the politicization of mourning.

Our aim in this chapter is to discuss "Making Loss the Centre," in which we bring together otherwise disparate ideas about environmental loss, and to consider the role of the podcast as a sounded ecology (Samuels et al. 2010) in part to disrupt aspects of the environmental movement that are frozen in patterns of response and ignore the importance and power of mourning. In the following pages, we draw from the work of Sigmund Freud and Judith Butler to explore contemporary conceptions of loss and mourning as they relate to the worldwide environmental destruction we currently face. Placed alongside understandings of the importance of loss, grief, and mourning as acts of environmentalist resistance, we consider how and why we have become stuck in destructive patterns of consumption and loss, and how we might break free (Clewell 2004).

Our presentation is threefold. First, we describe podcasts in general, and our podcasting practice with *CoHearence* in particular, as useful for exploring environmental loss and mourning. Second, teasing out connections from our episode, we offer a typology of environmental loss. Finally, drawing on the work of contributors to the episode, we expand on what Clifton Spargo (2004) calls "resistant mourning," a political response to loss in which the ethical relation to the lost Other does not end with the death of an individual, the extinction of a species, or the radical alteration of a landscape (Soper 2012, P2, 30:00–30:25; cf. Anderson 2007, 195; Riaño-Alcalá 2006; Smith 2001; Spargo 2001). Instead, the examples of resistant mourning we trace provide what Mick Smith would call "an opening through which the 'other' is made manifest in its irreducible difference to ourselves" (Smith 2001, 262) by challenging Freudian conceptions of mourning as a substitutive practice in which the lost object can be replaced. By refusing to accept that bodies (human and non), things, and places are replaceable, resistant mourning provides both ethical and political opportunities to "expand discursive spaces to include bodies that are not mourned in dominant discourse" (Cunsolo chapter 7). In this way, environmental mourning reconstitutes the nature of grievability, the differential allocation of the ability to publicly acknowledge the loss of a particular subject (Butler 2004, xiv) in radical and posthumanist ways (Eng and Kazanjian 2003, 8; Alaimo 2010; Haraway 2008; Ryan 2013; Butler 2004; Stanescu 2012).

## Podcasting

SUSAN MOORE: There is a very important role of language and art in relation to [environmental] loss.

MINUTE 38:25, "MAKING LOSS THE CENTRE" PART 2, *COHEARENCE*

Podcasts are on-demand audio files that are streamed or downloaded from the Internet to play on a computer or on a portable device at a user's convenience. While they have a great deal in common with audio documentaries, radio programs, soundscapes, audio theatre, and live performances, it is primarily through their relation to the Internet that podcasts are defined. The focus, length, and audio quality of podcasts vary, but most are available to the public for free. Episodes in the *Co-Hearence* podcast series are generally an hour long and appeal to those with an interest in environmental issues. While our resources are limited, we are concerned with aural aesthetics and aim to create podcasts that are not only informative or thought provoking, but are also pleasurable to listen to. As of April 2015 the two halves of the singular episode that is the focus of this paper have been accessed, downloaded, and/or listened to approximately five hundred times each since they were released in 2012.

Podcasts also offer an opportunity for sustained listening practice. *CoHearence* presents listeners with an informal and intimate portrait of the people we interview. The manner in which our contributors speak to us differs significantly from their lecture style or written practices, and listeners attune to the sounded "diction, phrasing, intonation, and emphasis" (Feld 2012, xvii) of their voices. The texture, timing, tuning, and timbre of voice can weigh more in dialogue (Keil and Feld 1994) than in the symbolic meaning of text (Turino 2008), which requires decoding and imagined intonation. Listeners can instantly project their own understandings onto verbal subtleties and the phenomenological experience of sounded information changes acoustemological knowing of place as we listen (Keil and Feld 1994; Feld 1996). This is particularly relevant for those who enjoy music or radio while engaging in other activities, where physical spaces become re-signified and associated with the acoustic environment of the moment. In some measure, and resonating with Krause's opening piece, by using an aural medium we acknowledge the essential component of sound for how we understand environment and loss (Krause chapter 1; Krause 2012, 2013; Feld 2012; Moten 2002).

In addition to practical considerations, our practice is based within a theoretical understanding of knowledge production and memorialization because podcasts represent a kind of archive. As documentarians of a moment in environmental thought that locates and dwells in loss, we embrace elements of German philosoper Walter Benjamin's historical materialist approach to memorialization and archiving (interpreted by Eng and Kazanjian 2003). Benjamin, a critical theorist and philosopher, distrusted hopeful and mono-linear romantic narratives of the past as well as popular Marxist fantasies of limitless progress in the worker's revolution. Eng and Kazanjian define Benjamin's particular version of historical materialism by contrasting it with historicism, the political production of history through the tactics of narrative and empathy to produce romantic and hopeful visions of the past and future. As a direct result of our medium, our narratives involve mosaics of voices and perspectives that are from separate pasts and individuals – non-synchronous and non-linear in time and space at the original moment of documentation and later in production: they disrupt the ease of romance and progress.

With narrative historicism, loss and nostalgia are unfortunate but necessary parts of progress (Smith 2001): "the concept of 'progress' and the concept of 'period of decline' are two sides of the same thing" (Benjamin 1989, 48). Empathetic historicism fails to fully engage the agency of the lost Other in its difference but instead wallows in melancholy (Smith 2001). Set against historicism, Benjamin promotes historical materialism, in which grand narratives of progress or decline are abandoned in favour of an understanding of historical subjects (animals for example) as much more than vessels for empathy. His historical materialist approach to loss insists that recorded history – memorialization – is not a matter of fate but rather the consequence of politics, multiple perspectives, subjects, victors, and actors. Within this more mobile framework, history and memory dialogically offer opportunities for ethical consideration instead of pity or inevitable naturalized placement (Cronon 1992).

## Talking Loss: Contributors to "Making Loss the Centre"

> PETER TIMMERMAN: One has to see loss as an integral part of human life; one has to see [loss] as one of the characteristics of who we ourselves are, and not to look away from it. [Environmental activists need] to use loss as a way of understanding our situation in more detail ... to try to ground the things we want to do more realistically.
>
> MINUTE 38:20, "MAKING LOSS THE CENTRE" PART 2, *COHEARENCE*

In collecting material for "Making Loss the Centre," we did not start with a list of individuals to interview. Instead we followed the threads of conversation with our interviewees to other contributors that could also speak to the relations between mourning and melancholy. This methodology was initially unruly, and it took us over eight months to mentally, cognitively, and emotionally digest the interviews before we could meaningfully organize them.

We began by interviewing Peter Timmerman and Catriona Sandilands, faculty members in the Faculty of Environmental Studies (FES) at York University. Peter Timmerman's interest in environmental philosophy and the work of Joanna Macy and Catriona Sandilands's work in queer ecologies and ecopolitics provided us with the historical and political grounding for our episode (Timmerman 2012, P1, 02:55–16:50; Sandilands 2012, P1, 02:55–16:50). Following Catriona Sandilands's advice on forming a better understanding of psychoanalytical perspectives on loss, we sought out the expertise of Susan Moore, an FES contract faculty and psychoanalyst (Moore 2012, P1, 17:05–24:13). We also spoke to Ella Soper, a postdoctoral fellow in FES and contract faculty member in the Department of English at the University of Toronto's Mississauga Campus, about the role of literature and art in resisting substitutive mourning practices (Soper 2012, P2, 24:00–36:20). For his thoughts on the importance of spirituality and mourning for the environmental movement, we sought out Ralph Carl Wushke, an ecotheologian, the minister at Bathurst United Church in Toronto, and a chaplain at the University of Toronto (Wushke 2012, P2, 07:20–12:50). Finally, Honor Ford-Smith, an FES faculty member and performance artist, told us about her work with marginalized communities in Toronto and Jamaica surrounding public commemorative performance that deals with profound personal and systemic violence, the criminalization of black youth, communal mourning, and reparation (Ford-Smith 2012, P2, 41:50–59:50). Each of these six individuals approached environmental loss in a different way, but all called for a politicization of mourning by focusing on particular losses, whether human or more-than-human, living or non-living, through ritual, creative, and artistic practices.

## Melancholy, Mourning, and Environmental Thought: Making Loss the Centre, *CoHearence* Episodes I and II

RALPH CARL WUSHKE: For a lot of environmental activists there is a frenetic, desperate sort of activism, in the sense that we've got to act – the earth is going to disappear [and] the planet as we know it is going to be gone – we've got to do something. So there is this intense desperation really ... But one of the things that I've heard when I've talked to people who are professional environmentalists, whose job it is to improve sustainability or recycling, is that there is this sense of despondency, that we're never going to get there, that it's too little too late. So that to me is where a ritual of lament and sitting with the loss so that you actually do the grieving that you need to do [is important].

MINUTE 10:50, "MAKING LOSS THE CENTRE" PART 2, *COHEARENCE*

We design *CoHearence* episodes using an investigative narrative structure and rely on a large topic to create a central theme to which expert individuals can speak. For "Making Loss the Centre," we asked our contributors a set of open-ended questions: what does melancholy and mourning have to do with the environmental movement and, as educators, how do we help students deal with environmental loss? In reorganizing fragmented sound clips from our interviews, we created an acoustic narrative with the aim of evaporating the messy interpretive role of the editors by replacing it with narrators who journey along with listeners to explore the interviews and content of the episode. After the journey, listeners are left with a mosaic of ideas and sounds that relate *kaleidophonically*[3] (Tracey 1970). Some connections are left for the listener to discover, and some are made more explicit.

Rather than describe the incremental details of "Making Loss the Centre" here – a redundant task when you might just listen to it yourself – we offer a reflection on how we see our interviewees' ideas connected now, in conversation with the research we have conducted since creating the episode. Our aim is not to attempt a comprehensive view of environmental loss, but to re-represent the expansive boundaries of our actual episode. We confine our discussion precisely to the ideas and themes offered by our contributors even as we are tempted to move beyond them. The typology of loss, presented in list form below, arises from moments of unconscious synchronistic and kaleidophonic consensus between our six contributors, and many of the concepts we have highlighted here bleed into one another. These connections are based in larger bodies of literature, but collecting them and presenting them as a list provides opportunities. For instance, the list is both a resource for future research to amend and expand in a practical fashion and a way to highlight relations among academic scholarship, creative response, environmentalism, and environmental loss that might otherwise remain hidden.

### 1) Theorizing environmental mourning requires expanding frameworks of individual-patient-focused psychic structures to encompass ideas of larger social and environmental consciousness

PETER TIMMERMAN: We know that we are going to have losses, we've already had losses, we're losing species everyday, and that this is something that human beings bear responsibility for. And yet, simultaneously, many activists ... are really in despair ... about what to do and about the situation where we will be able to do things. We will be able to recycle. We will be able to engage in various important activities … locally, regionally, and globally, and yet the overall momentum of what's going on is essentially going to doom large numbers of species and whole parts of the planet.

MINUTE 05:35, "MAKING LOSS THE CENTRE" PART 2, *COHEARENCE*

From the Ancient Greeks to current epistemological frameworks, it appears to us that the story of melancholy and mourning in Western culture is well documented (Radden 2002; Smith 2001; Comay 2005). In Freudian psychoanalysis, especially Freud's essay "Mourning and Melancholia" (1917), normal mourning is characterized by an individual's ability to move on or recover from losses by finding new attachments (Sandilands 2012, P1, 06:57–08:25; Timmerman 2012, P1, 08:27–09:10). If an individual fails to let go of a loss but instead finds that "a piece of the beloved is … lodged in [their] psychic structure" (Sandilands 2012, P1, 08:10–08:25), then Freud might have described their situation as pathologically melancholic. In his 1923 work, *The Ego and the Id*, Freud came to understand melancholy as not only a pathological response to the fear of death, but also as a part of the mourning process (86–8). In our podcast, Catriona Sandilands explained that Freud's view shifted from a conceptualization of melancholy as "an incomplete [or] perverted form of mourning" to an understanding that individuals must "incorporate the object into [their] psychic structure, leaving the mark of the beloved in [their] ego identification before [they] can let it go" (Sandilands 2012, P1, 12:10–13:00; see Freud 1947, 86–7). Peter Timmerman elaborated further, explaining that as Freud's thinking on melancholy progressed, he came to understand loss as constitutive of the nature of the self so that "the selves that we have, in themselves, are a kind of residue of all of this mourning" (Timmerman 2012, P1, 11:01–11:16; Clewell 2004, 61; Freud 1947, 35–6). "Melancholics are in mourning, then, but in a particular way" (Clewell 2004, 59), because the mourner refuses to move on from the loss, ultimately ruminating on the portions of the self that are made up of losses, internally angry at the portions of the self that are gone and may have abandoned the rest of the self.

While Freudian psychoanalysis is generally understood as relevant to personal losses, the episode draws on the work of Judith Butler (2004) to push the limits of this framework so that we can use it as a way to interrogate larger losses and patterns of loss, from the loss of individual species (Soper 2012, P2, 31:30–32:49), to the destruction of places (Sandilands 2012, P1, 29:25–33:20), to violence through global restructuring (Ford-Smith 2012, P2, 41:51–59:50). Butler's assertion that "some lives are grievable, and others are not" (Butler 2004, xvi) is central to a consideration of how environmental loss is understood in the contemporary context. In her explanation of Butler's notion of grievability, Catriona Sandilands told us that "loss ... and mourning [have] a social context and one of the important considerations is that we as a society find certain objects, what [Butler] would call grievable. We are allowed to mourn certain kinds of losses; there are rituals to support that. There are conditions in which we are allowed and encouraged to experience a loss as a loss" (Sandilands 2012, P1, 13:30–14:15). In *Precarious Life* Butler contends that there are two main forms of normative power that foreclose an ethical response to the perceived Other and render them ungrievable: first is the symbolic representation of the Other as inhuman; second is the radical effacement of the image of the Other, such that the image of the Other, and their loss, is excluded entirely from view. Grievability, Butler maintains, is often "circumscribed and produced in ... acts of permissible and celebrated public grief" and can operate "in tandem with a prohibition on the public grieving of others' lives" (37).

Although her critique is securely rooted in a consideration of the human subject, an extension of Butler's ideas into our discussion of environmental ethics and environmental loss in "Making Loss the Centre" seems a small but extremely productive step. On the one hand, environmental losses are not considered legitimate or grievable because, broadly speaking, the environment tends to be imagined as made up of natural resources available for human consumption and exploitation. On the other hand, there is a failure or refusal to represent, in a meaningful way, the loss of species, places, or whole ecologies. In Butler's words, "the first is an effacement through occlusion and the second is an effacement through representation itself" (147). As such, environmental losses do not count as grievable losses and have few socially accepted rituals to support them. Individuals are not encouraged to consider how the loss of more-than-human beings, entire species, and/or places might constitute the self. Further, we consistently fail to consider how the loss of the *relation* to the more-than-human Other might require mourning, as the removal of the "relational bind" itself fundamentally changes who we are (46).

Likewise, there are few rituals to support collective mourning for the environment, or to consider the ways in which environmental losses leave their mark on the psychic structure of a community, culture, or society. In our podcast, we explore the ideas of Freud and Butler as useful for thinking about not just individual psychic structure but also as important for understanding (Western) societal psychic structure as well, a topic we will return to below.

### 2) Our losses are systemic, and can be difficult to grasp or articulate: Their scale confounds us

HONOR FORD-SMITH: [Reparation] has to be not just a personal act ... it has to be a way of socially recognizing that these problems are not caused just by somebody willfully deciding to become a criminal ... these have a complex geology of social production.

MINUTE 49:05, "MAKING LOSS THE CENTRE" PART 2, *COHEARENCE*

The very concept of environmental loss is elusive. How can we describe environmental losses that encompass experiences as disparate and as intangible as the fear and risk of drinking straight from almost any fresh body of water, the postmortem encounter with a long-extinct species through literature or art, or the slow violence (Nixon 2011) of toxic industrial practices that have prolonged but devastating effects on life forms? In its ability to cover vast expanses of geographical space and geological time, as well as to differentially affect communities or to remain hidden in our narratives of progress and decline, environmental loss pushes the limits of our conceptual frameworks and renders our accepted mourning rituals, such as the funeral or romantic elegy, inadequate (see Menning, chapter 2). However, as Cunsolo discusses in chapter 7, humans react emotionally and physically to environmental losses, even if these losses are unknown.

While humans have been altering landscapes and wiping out species for some time (Marsh 1965), the current speed, consequence, and magnitude of our actions often defy comprehension. Scholars such as Rob Nixon (2011) and Ursula Heise (2008) have described some of the many specific ways in which large-scale and long-term environmental destruction and violence is hidden from view in the contemporary context. In particular, Nixon's "slow violence," the unspectacular devastation done to particular communities of people and environments over long periods of time, highlights the failure of our imaginative frameworks to cope with the temporal and geographical scale and complexity of environmental violence and loss. In many instances, environmental losses are only recognizable upon recourse to investigative technologies or novel imaginative frameworks. Each of the contributors to our episode, along with Nixon and Heise, point to the representational challenges of environmentalism and argue that the humanities in general, and art and literature in particular, are profoundly important for reimagining the pressing environmental issues of our time. The task of activist artists and writers is to provide new imaginative tools for confronting the issues of scale, time, and responsibility inherent in environmental loss. Nixon argues that "to intervene representationally entails devising iconic symbols that embody amorphous calamities as well as narrative forms that infuse those symbols with dramatic urgency" (10). While Nixon's work focuses on the ability of artists and writers to expose, confront, and reimagine the slow violence of environmental destruction and abuse, it is only a small step to extend this argument into discussions of environmental mourning. In our episode, Ralph Carl Wushke and Catriona Sandilands pointed to the rise of AIDS activism and elegy in the 1990s, particularly the iconic AIDS Quilt, as a form of politicized mourning that changed public understanding of HIV/AIDS. Aesthetic responses to AIDS allowed activists to publically and powerfully engage in politicized mourning and to establish the grievability of individual queer bodies by "acknowledge[ing] that this life, *this* life, is gone, this life has been snuffed out" (Sandilands 2012, P2, 13:27–14:00). As Ralph Carl Wushke suggests, an analogous approach to environmental mourning would both include the particularization and politicization of environmental losses, or the making of iconic symbols and new narratives of mourning, and allow for people to grieve those losses in a meaningful and reparative way (Wushke 2012, P2, 15:50–16:00).

In our episode, Ella Soper describes the particular power of literature and story to make otherwise inconceivable loss understandable in new ways: "story can take the enormity of ongoing crisis, like species extinction, and bring it back to the plight of the one, the individual, to make it manageable" (Soper 2012, P2, 32:20–32:50). We also see our podcasting practice as a part of a representational intervention, one in which sound, voice, academic theory, and conversational narratives come together to present new ways of thinking through environmental loss and mourning. Like this edited collection, our podcasts aim to challenge the ways in which we apprehend environmental loss by troubling and expanding notions of what exactly counts as (grievable) loss. We position our work in line with Nixon's (2011) examination of the "political, imaginative, and strategic role of activist-writers" (15) to create accessible, resistant, and iconic narratives of slow violence. Translating environmental losses into meaningful articulations of mourning presents a real challenge, but it is an action that we think is central to environmental studies at the most elemental level.

### 3) Mourning for that which we have never experienced, known, or encountered is difficult but necessary

ELLA SOPER: [With reference to Fred Bodsworth's *The Last of the Curlews*]
I think it's really instructive to be reminded that the tragedy we are talking of here falls solely within the economy of the human. Because the animal itself ... can't possibly know that it is the last of its kind ... So it's a form of dramatic irony ... that we as readers know something the curlew can't possibly know about his own fate, but [we should] recognize that [while] it's no less a tragedy for the species, [the idea of] tragedy itself is a concept we bring to bear in our understanding of what's happening in the world.

MINUTE 33:20, "MAKING LOSS THE CENTRE" PART 1, *COHEARENCE*

In some instances, environmental losses are discoveries of things that, for some of us, never existed in the first place. Whether we read a story about post-natural animals or extinct species (Soper 2012, P2, 28:30–29:17) or get the sense that we have lost a cultural practice or way of life, it is difficult to understand how environmental losses constitute both individuals and communities. At the individual level, Butler asserts that public displays of grief are complex but profoundly important parts of defining who we are: "I am as much constituted by those I do grieve for as by those whose deaths I disavow" (46). Public mourning also has a role to play beyond the individual, and helps to define the limits of the public itself: "The public is predicated on the exclusion of certain images, names, losses, and enactments of violence" (ibid., 38). So when environmental losses are addressed from the "very solid position" of our own modernity, "we reinforce our ability to destroy in the same moment that we mourn the fact that we have done so" (Sandilands 2012, P1, 29:30–32:36). The question of how environmental loss constitutes us can be suppressed or actively unaddressed.

Students of the environment discover just how much we have already lost through moments of revelation, sometimes grieving for species that have been eradicated altogether, such as the now extinct northern curlew (Bodsworth 1963), and sometimes reckoning the ongoing violence of an environmental wound that will not close (Soper 2012, P2, 33:20–34:02). We grieve, as Krause explains in chapter 1 of this collection, for the complete loss of acoustic space free from anthrophony, the persistence of human sounds in every corner of the planet (Krause 2012). We also grieve to discover histories, of colonization, mercantilism, and nation building in which loss, violence, and oppression are linked in troubling ways (Ford-Smith 2012, P2, 42:32–49:57). But in the absence of rituals to secure the grievability of the environment and more-than-human beings, confronting their loss, and how that loss is linked to human violence, oppression, and responsibility, is a representational challenge.

### 4) Some losses are not socially acceptable to admit or address, and are often actively unimagined

CATRIONA SANDILANDS: In a society in which homosexual attachments are not considered legitimate in the first place, it's not just that we have no rituals surrounding their loss, it's that we declare that they are really ungreivable losses – they weren't really there in the first place so how can we possibly acknowledge that they were losses?

MINUTE 15:51, "MAKING LOSS THE CENTRE" PART 1, *COHEARENCE*

Loss may be easily identifiable but not socially acceptable to acknowledge or incomprehensible to some as matters of grief (Fowlkes 1990; Kristeva 1989; Butler 2004, xvi, 33). By linking the active disavowal of the subjectivity of particular communities of human beings directly to the intentional and powerful ways in which the inherent value of the environment and more-than-human beings are denied, Catriona Sandilands extends Butler's notion of grievability beyond the realm of the human and into the more-than-human world: "what would it mean to mourn the loss of the fossil fuels? It's inconceivable, to mourn the gas that we've burned. It's just absolutely impossible. If we are going to continue to use it, it has to be replaceable. We cannot acknowledge that the thing simply does not exist anymore" (Sandilands 2012, P1, 34:47). As the more-than-human world falls outside of our common understanding of subjectivity, it can be "actively unimagined" (Nixon 2011) as a subject in the same way that women, queer people, people of colour, or poor people have been dehumanized and delegitimized through colonialist and patriarchal systems and discourses (Cunsolo chapter 7). Further, the more-than-human world often falls within the economic relations of late capitalist society and, as such, must remain ungrievable so that it can be used as a resource without consideration for its irreplaceable intrinsic value. The movement from things and people – valuable in their own right, though already problematic in their initial (human) objectification – to commodity/resource is important for environmental concerns about mourning.

Some of these ungrievable situations are not as difficult to contemplate as mourning liters of gas. For instance, James Stanescu (2012) explores Butler's "precarious life," environmental loss, and the political in relation to animals: "mourning is stitched to the questions of what and who gets to count as human, and, therefore, to the very questions of how we constitute the political" (569). Stanescu considers the ways in which animal rights activists are expected to disavow their grief at the loss of animal life that confronts them in public spaces: "those of us who value the lives of other animals live in a strange, parallel world to that of other people ... most people's response is that we need therapy, or that we can't be sincere" (568); "[deli meats] are the 'nameless and faceless deaths' that come to constitute our world" (579).

In Stanescu's example, animal rights activists interpret animals as intrinsically valuable Others consumed by humans who deny their subjectivity. While animal flesh may be actively unimagined (Nixon 2011) as animal lives on the dinner table, obscured by industrial farming, butchering, processing, and marketing techniques, animal rights activists ask us to consider the ethics of acknowledging animal agency to reckon a reparative stance. Conceptualizing the consumption of animal flesh as an example of environmental loss may be more straightforward than thinking about the grievability of fossil fuels, but both examples bring attention to the capacity for more-than-human subjectivity. For humans, speaking of and giving voice to loss can be difficult, involving compounding barriers to articulation of loss, as people who are currently grieving for and experiencing environmental losses firsthand are often already marginalized from power centres and are hegemonically unimagined (Cunsolo chapter 7).

Even while working to imagine environmental losses as grievable, there are social conventions dictating "social responsibility" to keep one's grief hidden (Ferguson 1973). Social norms dictate not only what constitutes a legitimate source of grief, but also dictate the time and space, the parameters in which expression of grief is allowed. Having publicly admitted the depth of her own compassion for a tree on television, ecologist Phyllis Windle describes her embarrassment at acknowledging this emotional attachment to the health of her research subjects, our planet, and its species in such a public fashion (1992). Even in the face of her own embarrassment and the threat of public ridicule, Windle ultimately concludes that we stand to lose significantly more by failing to grieve publicly – or failing to talk about our grief – than we do by grieving. By grieving publicly, we can help reshape and enlarge understandings of grievability itself.

### 5) The loss of the ability to conceive of a future

PETER TIMMERMAN: Through the Cold War, this possibility that we could obliterate everything generated various forms of mourning and melancholia in the political domain, but also in the activist domain: would we ever get out of this situation, and so on.

MINUTE 22:52, "MAKING LOSS THE CENTRE" PART 1, *COHEARENCE*

The invention of the atomic bomb and the realization that human industrialization is causing rapid climate change threaten our ability to have concrete faith in the future. In our impulse to improve, we have acted on ideas and desires whose consequences will echo for generations, such as mega-dams and the introduction of invasive species. These failed dreams cannot capture their original utopian aims and represent a melancholic state in their making and conception. Our technologies are outdated in the moment of their creation, and they linger with toxic consequences, refusing to leave us, reminding us of our hubris, and channeling our anger back at ourselves. These ruined projects, broken and fragmented metaphors for a better future, give rise to a unique form of melancholic detachment (Timmerman 2012, P1, 24:25–26:25). The existence of these dangerous projects as birth rights to new generations now represent a kind of *given* melancholic inheritance.

In addition to a general sense of loss and anxiety at the failed projects of modernity, we encounter anticipatory grief as we witness the decline of particular species, the continuation of extractive and destructive approaches to the environment, and events that doom entire populations or ecologies (Timmerman 2012, P2, 05:35–4:00; Cunsolo chapter 7). Our feelings of being powerless in the face of an impending loss alongside the ungrievability of more-than-human subjects can be destructive and debilitating, particularly for communities most effected by environmental degradation and violence. As Jessica Marion Barr asks in chapter 8, how can we imagine a resistant form of anticipatory mourning that does justice to the thing not yet lost while at the same time challenging the derealization of environmental loss itself? Even further, how can we imagine a response to wounds that have yet to be opened or have always existed in our lives?

Our experiences of past losses allow us to imagine and project future losses, but perpetual anticipatory anxiety can turn into fetish. Ella Soper wonders what is implied in our tendency to transmute the idea of apocalypse as "one world giving way to another" to instead "talk about apocalypse in contemporary parlance [as] the end of ends ... there is no world beyond it" (Soper 2012, P2, 34:20–35:09). While phobias of germs or fears of conspiracies may have once been more easily dismissed as the works of an over-wrought mind, today fears of rising sea levels, super bugs, and a toxic day-to-day reality are legitimate realities and worries that evade dismissal.

### 6) Loss and consumerism are linked

SUSAN MOORE: If we think of our culture [as] a melancholic culture, we [can see that we are] not willing to work through a loss, [we are] still attached to some original lost object; I think that rather than working through that loss what happens is a more manic kind of response ... rather than looking at what we have lost, and loss itself ... we try to replace that loss by filling it with consumer satisfactions and replace it with new things.

MINUTE 23:20, "MAKING LOSS THE CENTRE" PART 1, *COHEARENCE*

A key component of mourning in the Freudian psychoanalytic model is the possibility of reattachment. Clewell (2004) writes that Freud eventually retracted his notion of overcoming loss through replacement. Freud's original formulation is both too simple and overly narcissistic. However, his later revisions on reattachment still place self-interest as an important motivator for experiencing grief. Freud's eventual idea is that the self is an accumulation of losses (ibid.). Clifton Spargo (2004) points out that Freudian substitutive mourning, which has become "interwoven with other historical, religious, and philosophical conceptions of grief" (21), has much in common with our "larger cultural preference for utilitarianism ... in which any thought for the other can be distilled to a thought or care for the self" (ibid.). There are trends towards modern mourning practices that resist reattachment and strive towards "the art of losing" (Ramazani 1994); however, people continue to purchase commodities in order to ameliorate the anguish of loss through what is popularly called "retail therapy" (Plastow 2012; Garg and Lerner 2013). This is an important aspect of loss and environmental justice today, particularly as our losses accelerate; those consuming the most are proportionately losing the least while the inverse is also true (Sandberg and Sandberg 2010).

The psychic structure of mourning and commodity fetishism points towards disconcerting linkages between lost objects, species, people, spaces, and the desire to replace them with new commodities (Moore 2012, P1, 23:18–24:15; Comay 2005). This drive to replace, however, requires us to sever our ethical relation with the Other. In so doing, we fail in our ethical responsibility to the Other and run the risk of avoiding or forgetting what we have lost and how we lost it in the first place.

Not only do we consume material environmental objects, but environmental campaigns are also staged as commodities, moving activists from the consumption of, or concern for, an endless series of disaster and loss products. Environmental campaigns that utilize capitalist marketing find and retain modern customers and consumers. These campaigns implement the same tactics as corporate advertisers:

> We develop intense attachments to things, and ... once that object is gone, we then develop an intense attachment to another thing. And I think that's kind of how media campaigns around environmental destruction take place; well, first it was this forest, and now it's this forest, and first it was this species, but now it's this species ... we get upset and we protest the loss, but we do so from the very solid position that we have already come to the point that we are absolutely capable of conquering [nature], so we reinforce our ability to destroy in the same moment that we mourn the fact that we have done so. But we don't really sit with that loss, or understand how we might be constituted by that loss. We look at it and we take a picture of it and we put it in National Geographic and we say "isn't it sad that this culture is on its way out" or that "this species is on its way out," but we don't understand how our own modernity, our own psychic being in the present is in fact constituted and reconstituted and reconstituted by that loss. (Sandilands 2012, P1, 28:37)

In an uncomfortable symbiosis, environmental organizations find new sites for action as new resources are discovered, arising in response to the expansion of our losses. In a perverse sense, capitalism is feeding the growth of the environmental movement. Things are lost to our progress (Smith 2001), which we grieve, but only as a way of recounting stories of tragic-but-necessary sacrifices for our advances (Cronon 1992). For example, the industrial revolution is often framed as an evil that was required to create things like unions, workers rights, and environmental reform. Must we lose the planet in order to create a narrative that allows us to save it?

Bruce Braun's description of the fight for Clayoquot Sound in British Columbia encapsulates how we confirm our modernity, our moral and social progress, by mourning loss (2002). For Braun the "binary logic" (2) of most environmental movements is "predicated on the idea of nature's externality" (ix) and hides the ways that "natural" landscapes are "complex terrains of culture, politics, and power" (x). Further, environmental movements that understand nature narrowly, as the opposite of society, produce environmental activism that is "complicit in [the] forms of erasure and abjection" that characterize the modern industrial complex and that have led to the losses for which we mourn in the first place (2). By focusing too narrowly on "natural" losses, environmentalists can fail to keep sight of the "unnatural" structures that have already produced, altered, and now govern the future of our environment in the Anthropocene.

## Responding to Resist

ELLA SOPER: Can we move [toward] a more ethical engagement with works of literature: a more ethical engagement with the plight of others that doesn't seem to other them or to reassert the comfortable binary between self and Other? ... [Examples of resistant mourning in literature] can really problematize that [binary] to interesting and perhaps rather alarming effect, as posthumanist studies tend to do. You know, really blurring the boundaries between, for example, what is human and what is non-human, what is self and what is Other?

MINUTE 26:30, "MAKING LOSS THE CENTRE" PART 2, *COHEARENCE*

While guiding our listeners through the types of environmental loss explored above, we uncovered moments in our episodes that presented opportunities to respond to these problems of loss in meaningful and radical ways. We also came to see our own podcasting practice as a particular response to environmental loss. Typically, when confronted with problems environmentalists are encouraged – and encourage others – to look for solutions. There is a continual tension between the need for an immediate response to environmental catastrophe and meaningful long-term reparative strategies. However, scenarios in which new technological innovation and awareness campaigns overshadow the possibility of foundational changes in behaviour or creative reimaginings of our relationship with the environment reflect a kind of fetishized new attachment. Instead of protesting, instead of changing the world, how might we first try to comprehend how we are constituted by our losses (Sandilands 2012, P1, 29:25–33:20; Butler 2004; Eng and Kazanjian 2002)? How might we sit with a loss, embrace melancholia, and mourn in a way that prevents expedient reattachment?

In *The Ethics of Mourning*, Spargo (2004) describes the particular power of "resistant mourning" (10) to trouble our understanding of socially accepted responses to loss. For Spargo, resistant mourning is "a strain of mourning, in which there is opposition to psychological resolution and to the status quo of cultural memory" (6). In the refusal to conclude mourning according to substitutive practices, which fail to maintain an ethical relationship with the lost Other, Spargo suggests that "unresolved mourning becomes a dissenting act" (ibid.) that pushes against the limits of what is grievable and how we grieve because "mourning, especially in its most extreme cases, brings our assumptions about reality into question" (21). While Spargo's text focuses on the specific ability of literature to engage in resistant mourning, the concept is important for environmentalism more broadly. As environmentalists continue to struggle to substantiate environmental losses as legitimately grievable, as Jessica Marion Barr's contribution to this collection explains, resistant mourning practices offer practical responses to loss that raise political and ethical considerations (chapter 8). By tarrying with grief we improve our creative capacities to respond to loss, and we deepen our rituals and bonds to community, forging the support, strength, and determination needed to resist further losses. Below we highlight some of the resistant mourning practices described by contributors to "Making Loss the Centre."

### 1) We can respond creatively to loss, and this helps

HONOR FORD-SMITH: It can never be perfect. It can never completely regain, but it is a way of renewing ... The elegiac impulse ... the lament ... make the past present in a way that aims to be reparative in some way.

MINUTES 48:18–49:05, "MAKING LOSS THE CENTRE" PART 2, *COHEARENCE*

Melanie Klein (1940) argues that, following a loss, opportunities for reparation commonly involve a creative sublimation of negative feelings (Moore 2012, P1, 36:10–36:34); by creatively giving language to a loss, we might begin to understand how it constitutes us more fully (ibid., 22:10–23:30). Not only can creative responses present opportunities to explore unconventional approaches to addressing, managing, and growing from loss, but, with the ability to articulate loss, melancholic individuals and institutions can direct frustration and anger into description and away from the unidentifiable frustration with illusive lost portions of self.

Each person we interviewed for our episode echoed Klein's insistence on the importance of creatively engaging with loss. As resonating with Jessica Marion Barr, creative expressions of mourning in this collection, including literature and poetry, visual or performance art, dance, music, and digital media ensure that we maintain an ongoing relationship with what or who has been lost, and become political as we witness "that *this* life, *this* life, is gone, *this* life has been snuffed out" (Sandilands 2012, P2, 14:26–14:31). Our work with this episode of *CoHearence* represents precisely this kind of creative effort and responds directly to Butler's (2004) assertion (echoed by Nixon 2011) that work in the humanities has an important role to play in redefining the frame of what is or is not grievable. The elegiac impulses (Ford-Smith 2012, P2, 48:18) common to art and literature can serve as acts of protest that challenge the limits of our imaginative cultural frameworks for dealing with loss in an ethical and ongoing way. Creative acts of resistant mourning purposefully trouble the "comfortable binary between self and Other" (Soper 2012, P2, 26:50–26:55) and allow us to hold onto losses to allow an imaginative practice that involves ethical-political responses and environmental consciousness.

Benjamin argued for the importance of creative allegory for overcoming problems associated with inauthentic progress, where progress can only happen at the expense of nature or when green washing is marketed as progress, and anthropocentric empathy, where the moral burden for the Other is derived from either the polar separation of self and Other or total sublimation of each into one (Smith 2001; Timmerman 2012, P1, 24:25–26:25). Ideally, one should be able to recognize both the divisions and connections between self and Other simultaneously as co-constituents of reality. Allegorical work, as stories, art, and performance, in particular "offers redemption of a kind because it reopens the slightest gaps between the subject's representation of the object of loss and the object itself" (Smith 2001, 370) and makes painfully obvious our lack of pure connection with the Other. While the desire to have total empathy with an object assumes too much and representing the object as a narrative consequence of evolution is totalizing – where the destruction of nature is the regrettable necessity of modernity – creative allegorical works make attachment to the work (the object, the Other, nature) at once so obvious and so lacking that audiences find opportunities to struggle with their ethical positioning (Mark 2014), which for our purposes is their ethical obligation to the environment. Allegory gives audiences permission to come to personal conclusions about environmental issues without the persuasive mystique of metaphor. Allegory as art – or vice versa – presents an opportunity for dialogic commons through open and public interpretation, a dialogue that less gregarious and interpretive forms of communication about environmental problems can preclude.

### 2) Ritual helps

RALPH CARL WUSHKE: We are ritualizing beings, a ritualizing species. It's not just about static ritual; it's about our capacity to ritualize, which is our way of figuring out who we are, our way of ... shaping our understanding [and] also our reality. At a very basic level, when we ritualize we are in the process of creating the world that we want, or creating the world as we imagine or hope it would be. That's ethical ritual at least.

MINUTE 08:28, "MAKING LOSS THE CENTRE" PART 2, *COHEARENCE*

As a ritualizing species (Driver 1998), we have an opportunity to transform how we experience loss through substantive practices. Rituals, as Nancy Menning outlines in chapter 2, can provide chances to create and change patterns of behaviour, and they are the traditions that allow us to make sense of ourselves and of the world. Ethical ritual can allow us to practice how we would like the world to be and can be a type of resistant mourning. For example, Ralph Carl Wushke and his congregation use a breathing ritual to find their connection with all living things through time (2012, P2, 18:33–20:00), and Peter Timmerman suggests that Joanna Macy's ritualized despair and empowerment activities are particularly useful for groups of people confronting loss (Timmerman 2012, P1, 20:00–21:47; Macy 1991; Macy and Brown 1998). Memorials and memorialization also ritualize loss (Junge 1999; MEMO 2013). While rituals can encourage us to move forward, the movement can be into a circle or a spiral, and not in a linear fashion. Mourning rituals are in this model, a perennial activity, something to be repeated and practised. To lose is to be human after all, "[Citing Ecclesiastes:] There is a time for mourning, a time for laughing, a time for planting, and a time for harvesting, a time for being born, and a time for dying" (Wushke 2012, P2, 09:35–09:50).

Creating rituals that help us to reimagine what can be grievable is fundamental to resistant mourning practices. In our episode, we highlight the profound importance of ritual and memorial in reckoning the particularity of individual losses (Sandilands 2012, P2, 13:22–14:22; Wushke 2012, P2, 15:01–16:00; Ford-Smith 2012, P2, 55:29–55:36). Thus the AIDS quilt, reading names aloud at the Holocaust Memorial, or the ghost bike memorials in Toronto become important political projects for ensuring the visibility of groups of people who have been lost and whose loss was previously unacknowledged or invisibilized (Butler 2004). Acts of memorialization that combine environmentalism with environmental justice concerns, such as Honor Ford-Smith's work in Toronto and Jamaica (described below) or Jessica Marion Barr's artistic renderings, reimagine the world such that environmental subjects are among the grievable.

### 3) Community is essential

HONOR FORD-SMITH: I've had instances where people refuse to participate in the work because they don't want to stir up these very unpleasant memories. I think one way [in answer to our question: "how do you do this work without succumbing to melancholy?"] is to build communities of mourning. This is not my phrase. This is the phrase of a woman called Pilar Riaño-Alcalá ... whose work has inspired, in a way, some of what I am doing.

MINUTE 58:59, "MAKING LOSS THE CENTRE" PART 2, *COHEARENCE*

One of our goals in our podcasting practice is to impact the digital commons (Di Battista and Mark 2013; Hawk 2011). With "Making Loss the Centre" we also aimed to gather a community of mourning such that we might foster a discussion of resistant mourning practices. In the widest possible sense we aimed with our first episodes to capture an audience that included members who were academic, public, and artistic. In theory, our communications are available to anyone with access to the Internet.

Our efforts also had practical significance. Following the release of the first half of "Making Loss the Centre," we held an aural "screening" with the podcast contributors from our department. The roundtable discussion that followed critiqued our initial approach to the topic as overly solutions-focused, but the mood also reflected an enhanced sense of interdisciplinary common purpose, language, and feeling. Participants felt particularly grateful for the opportunity to consider the topic together in an elevated dialogue. The common feeling went beyond mere collegiality, but included the admission that we in the room felt deep loss.

In addition, initiating our podcast series with such a topic created a space for both of us, the authors, to foster our own personal community of mourning. The benefits of creative response are in the process of creation as much as in the final product. The creative ritual of preparing our equipment, transcribing our interviews, passing back and forth ideas, recording our own voice-overs and music allowed us to generate our own findings about melancholy and mourning and develop a shared sense of mourning together for the discovery of this neglected discourse. Additionally, Ralph Carl Wushke's congregation was pleased to hear of his collaboration with *CoHearence*. Inevitably, creative responses to loss draw in widening circles of participants through collective exchange.

In her book *Dwellers of Memory* Pilar Riaño-Alcalá (2006) asserts that "memory becomes a strategic tool for human and cultural survival" (12) when violence and ongoing loss threaten the material and social well-being of a community. It is not surprising, then, that all of the contributors to "Making Loss the Centre" discussed the importance of memory and community to environmental mourning practices in some way. However, the most explicit example of resistant communities of mourning in the episode was drawn from Honor Ford-Smith's work.

There are climbing rates of homicide in Jamaica and the Jamaican diaspora due to global restructuring and liquid modernity (Ford-Smith 2011; Baubman 1998; Blackshaw 2005, 32–3). Honor Ford-Smith examines how contemporary violence reflects past experiences of colonial violence and marginalization in particular communities, echoing through memory. She is also profoundly concerned with the ways in which individual losses reflect systemic forms of colonial violence and globalization and her performative memorialization of violence in Toronto and Jamaica draws on the successes of the Madres de Plaza de Mayo (Ford-Smith 2012, P2, 50:15–52:40). The Madres, a group of mothers in Argentina who collectively staged vigils for their "disappeared" children during the Dirty War of the military dictatorship from 1976–83, created a ritualized resistant mourning and protest movement that gained international recognition. At these performances of public grief, the mothers publicly and strategically grieved for their lost children to shame the dictatorship. Mothers wore a picture of their child around their neck and a white scarf to signify their status during night marches to the presidential palace. The marches themselves displayed and created unity through muscular bonding and synchronous movement (McNeill 1995).

Outside the Eaton Centre in Toronto, Honor Ford-Smith helped organize a public performance piece that drew a found audience to witness the number of working-class youth deaths in 2005, "the proverbial 'Year of the Gun'" (Ford-Smith 2012, P2, 53:00–53:05). After following a coffin that displayed photographs of the deceased youth, audience members read letters from the dead and viewed other images that tied these deaths to global violence. Two years later, the organizers from Toronto transformed the work into a commemorative walk for mothers in Jamaica who have lost children to violence. The walk moved through gang territories and mothers carried pictures of their children, sang songs together, and wore a similar cloth (i.e. a scarf, sash, or shawl) to symbolize their collective grief (Ford-Smith 2011). Within the context of Ford-Smith's work, some community members and families of the deceased voiced concerns that, by collectivizing grief, it might be enhanced, and feed back into further violence and retribution in their communities (Ford-Smith 2012, P2, 46:46–47:19). Ford-Smith told us that, in working as an activist with communities confronting violence and loss, there are risks that must be considered when centring art and political actions around collective mourning: "and unless you have a community to work through it, [grief] becomes a difficult thing" (ibid. 47:19–47:26). In collectivizing mourning, there are risks one must consider against the benefits of a more unified voice.

## Conclusion

AMANDA DI BATTISTA: How can we engage in the politicization of mourning without becoming debilitated and re-traumatized?

MINUTE 58:40, "MAKING LOSS THE CENTRE" PART 2, *COHEARENCE*

ANDREW MARK: And there it was, the focus on community throughout this episode – the idea that the politicization of mourning is too big to do individually.

MINUTE 59:49, "MAKING LOSS THE CENTRE" PART 2, *COHEARENCE*

Producing *CoHearence* has given us insight into how a new medium can present fresh epistemological perspectives, but we can also recommend the practice as a creative activity and ritual for confronting environmental problems. Not only does our work facilitate community in interesting ways but the sounded connections we created in the episode also challenged disciplinary boundaries and enriched the stories we now tell about melancholy and mourning.

The knowledge we represent in our episodes was transmitted to us aurally; the in-person exchanges were both personal and professional and involved high-tech equipment, meeting times and places, travel, eye contact, hot beverages, and all of the accompanying micro-timed syntactic body language of human communication and proxemics (Hall 1966). In our written work, referencing and acknowledgement take on a whole new form(ality), intention, and meaning. The sensation is clinical in our exacting copy editing, with only masked residues of the feeling-full nature of our podcasts, complete with errors, colloquial aphorisms, run-on ideas, and our awkward newness to the medium.

Podcasts are made up of fragments of past conversations schizophonically (Schafer 1977) torn from their moment of utterance, and reorganized in a utopian fashion. These words are lost in a way, and as Eng and Kazanjian (2002) have outlined in their speculations on melancholia and semiotics, it could be that processes of loss are what constitute "domain[s] of remains" and "traces open to signification" (4). The architecture of our ability to communicate at all is built upon reference to lost places, groups, and individuals: things that may not be present. As we have demonstrated above, many of these memories are our own, but they are also social and exist beyond the individual (Ford-Smith 2011, 10; Creet 2010, 3–26; Connerton 1989; Ricoeur 1999). Further, our vulnerability drives us to communicate and is the root cause of our mourning, our melancholia, and our need for community (Stanescu 2012).

The form of the podcast itself, as a sounded ecology, allows for conversational flows, synergies, and connections between the voices of contributors, music, and additional recorded sound, creating a virtual space in which the affect of sound and voice adds to the listener's cognitive and emotional response to the content presented. This is a particularly productive quality for a discussion of environmental loss and resistant mourning, explored both in our podcast and in these pages, as attention to aural sounding presents an opportunity to imagine a listening practice that aids our environmental movement, a movement that is focused primarily on visual concerns (see Krause chapter 1; Krause 2012; 2013) and visual means of communication. By creating an audio document that requires listeners to pay attention to heard-resonances between ideas and sounds, our podcasts provide opportunities for new imaginative frameworks for dealing with environmental loss that might enable us to rethink our relation to the more-than-human world.

In writing, we may lose aural voice and its qualities, but we also gain new modes for experiencing the voices of our contributors. Our podcasting practice, in its relation with voices and episodes from the past, is ongoing, refreshed here through new references built upon old. This chapter represents another step in our personal mourning of environmental loses and an effort to publicly acknowledge our personal ongoing grief as others have done (Cunsolo chapter 7; Stanescu 2012; Windle 1992).

## NOTES

1 In this chapter we use many excerpts from our podcast series. Readers can locate quotes by referring to the timecode in the specific episode we reference. Our in-text citations of this work include timecodes instead of page numbers and P1 or P2 for Part 1 or 2 of the episode.

2 The authors acknowledge the generous support of the Network in Canadian History and Environment and the Faculty of Environmental Studies at York University in funding *CoHearence*.

3 Andrew Tracey (1970) uses the term kaleidophonic to describe the manner in which Zimbabwean mbira dzaVadzimu song can shift orientation or be multiperspectival in its organization, particularly when multiple instruments and voices join together. While the arrangements and sequence of notes remain the same, multiple frameworks of perception are simultaneously possible.

## INTERVIEWS

Ford-Smith, Honor. 2012. Interview in "Melancholy, Mourning, and Environmental Thought: Part 2 Making Loss the Centre," *CoHearence*, produced by Amanda Di Battista and Andrew Mark. Podcast. 8 March. www.niche-canada.org/cohearence.

Moore, Susan. 2012. Interview in "Melancholy, Mourning, and Environmental Thought: Part 1 and 2 Making Loss the Centre," *CoHearence*, produced by Amanda Di Battista and Andrew Mark. Podcast. 7 February and 8 March. www.niche-canada.org/cohearence.

Sandilands, Catriona. 2012. Interview in "Melancholy, Mourning, and Environmental Thought: Part 1 and 2 Making Loss the Centre," *CoHearence*, produced by Amanda Di Battista and Andrew Mark. Podcast. 7 February and 8 March. www.niche-canada.org/cohearence.

Soper, Ella. 2012. Interview in "Melancholy, Mourning, and Environmental Thought: Part 2 Making Loss the Centre," *CoHearence*, produced by Amanda Di Battista and Andrew Mark. Podcast. 8 March. www.niche-canada.org/cohearence.

Timmerman, Peter. 2012. Interview in "Melancholy, Mourning, and Environmental Thought: Part 1 and 2 Making Loss the Centre," *CoHearence*, produced by Amanda Di Battista and Andrew Mark. Podcast. 7 February and 8 March. www.niche-canada.org/cohearence.

Wushke, Ralph Carl. 2012. Interview in "Melancholy, Mourning, and Environmental Thought: Part 2 Making Loss the Centre," *CoHearence*, produced by Amanda Di Battista and Andrew Mark. Podcast. 8 March. www.niche-canada.org/cohearence.

## REFERENCES

Anderson, Eric. 2007. "Black Atlanta: An Ecosocial Approach to Narratives of the Atlanta Child Murders." *PMLA: The Modern Language Association of America* 122, no. 1: 194–209.

Bauman, Zygmunt. 1998. *Globalization: The Human Consequences*. New York: Columbia University Press.

Benjamin, Walter. 1999. "N [Re the Theory of Knowledge, Theory of Progress]." In *Benjamin: Philosophy, Aesthetics, History*, edited by Gary Smith. Chicago: University of Chicago Press.

Blackshaw, Tony. 2005. *Zygmunt Bauman*. New York: Routledge.

Bodsworth, Fred. 1963. *Last of the Curlews*. Toronto: McClelland and Stewart.

Braun, Bruce. 2002. *The Intemperate Rainforest: Nature, Culture and Power on Canada's West Coast*. Minneapolis: University of Minnesota Press.

Butler, Judith. 2004. *Precarious Life: The Powers of Mourning and Violence*. New York: Verso.

Clewell, Tammy. 2004. "Mourning beyond Melancholia: Freud's Psychoanalysis of Loss." *Journal of the American Psychoanalytic Association* 52, no. 1: 43–67.

Comay, Rebeca. 2005. "The Sickness of Tradition: Between Melancholia and Fetishism." In *Walter Benjamin and History*, edited by Andrew E. Benjamin, 88–101. New York: Continuum International Publishing.

Connerton, Paul. 1989. *How Societies Remember*. Cambridge: Cambridge University Press.

Creet, Julia. 2010. "Introduction." In *Memory and Migration: Multidisciplinary Approaches to Memory Studies*, edited by Julia Creet and Andreas Kitzmann, 3–26. Toronto: University of Toronto Press.

Cronon, William. 1992. "A Place for Stories: Nature, History, and Narrative." *Journal of American History* 78, no. 4: 1347–8.

Cunsolo Willox, Ashlee. 2012. "Climate Change as the Work of Mourning." *Ethics and Environment* 17, no. 2: 137–64.

Di Battista, Amanda, and Andrew Mark. 2013. "CoHearing Consequences: Podcasting, Ecopolitics, and the Power of Listening." In *Green Words/Green Worlds: Environmental Literatures and Politics*, edited by Catriona Sandilands and Ella Soper. Forthcoming.

– 2012. "Melancholy, Mourning, and Environmental Thought: Parts 1 and 2, Making Loss the Centre." *CoHearence*. Podcast. 7 February and 8 March. www.niche-canada.org/cohearence.

Driver, Tom F. 1998. *Liberating Rites: Understanding the Transformative Power of Ritual*. Boulder: Westview Press.

Eng, David L., and David Kazanjian, eds. 2002. *Loss: The Politics of Mourning*. Berkeley: University of California Press.

Feld, Steven. 1996. "Waterfalls of Song: An Acoustemology of Place Resounding in Bosavi Papua New Guinea." In *Senses of Place*, edited by Steven Feld and Keith Basso, 91–136. Seattle: Washington University Press.

– 2012. *Sound and Sentiment: Birds, Weeping, Poetics, and Song in Kaluli Expression*. Philadelphia: University of Pennsylvania Press.

Ferguson, Sarah. 1973. *A Guard Within*. New York: Pantheon Books.

Ford-Smith, Honor. 2011. "Local and Transnational Dialogues on Memory and Violence in Jamaica and Toronto: Staging *Letters from the Dead* among the Living." *Canadian Theatre Review* 148 (Fall): 10–17.

Fowlkes, Martha. 1990. "The Social Regulation of Grief." *Sociological Forum* 5, no. 4: 635–52.

Freud, Sigmund. 1947. *The Ego and the Id*. Translated by Joan Riviere. London: Hogarth Press. 4th edition.

Freud, Sigmund. 1957. "Mourning and Melancholia." In *The Standard Edition of the Complete Psychological Works of Sigmund Freud, Volume XIV (1914–1916): On the History of the Psycho-analytic Movement, Papers on Metapsychology and Other Works*, translated and edited by James Strachey, Anna Freud, Alex Strachey, and Alan Tyson, 237–58. London: The Hogarth Press and the Institute of Psychoanalysis.

Garg, Nitika, and Jennifer Lerner. 2013. "Sadness and Consumption." *Journal of Consumer Psychology* 23, no. 10: 106–13.

Hall, Edward. 1966. *The Hidden Dimension*. New York: Doubleday.

Haraway, Donna. 2008. *When Species Meet*. Minneapolis: University of Minneapolis Press.

Hawk, Byron. 2011. "Curating Ecologies, Circulating Musics: From the Public Sphere to Sphere Publics." In *Ecology, Writing Theory, and New Media: Writing Ecology*, edited by Sidney I. Dobrin, 160–79. New York: Routledge.

Heise, Ursula K. 2008. *Sense of Place and Sense of Planet: The Environmental Imagination of the Global*. New York: Oxford University Press.

Klein, Melanie. 1940. "Mourning and Its Relation to Manic-Depressive States." *International Journal of Psycho-Analysis* 21: 125–53.

Keil, Charles, and Steven Feld. 1994. *Music Grooves: Essays and Dialogues*. Chicago: University of Chicago Press.

Krause, Bernie. 2012. *The Great Animal Orchestra: Finding the Origins of Music in the World's Wild Places.* New York: Hatchette Book Group.

Kristeva, Julia. 1989. *Black Sun: Depression and Melancholia*. Translated by Leon Roudiez. New York: Columbia University Press.

Macy, Joanna. 1991. *Mutual Causality in Buddhism and General Systems Theory: The Dharma of Natural Systems*. Buffalo: State University of New York Press.

Macy, Joanna, and Molly Brown. 1998. *Coming Back to Life: Practices to Reconnect Our Lives, Our World*. Gabriola Island: New Society Publishers.

Mark, Andrew. 2014. "Refining Uranium: Bob Wiseman's Ecomusicological Puppetry." *Environmental Humanities* 4: 69–94.

Marsh, George Perkins. 1965. *Man and Nature*. Seattle: University of Washington Press.

McNeill, Walter. 1995. *Keeping Together in Time: Dance and Drill in Human History*. Cambridge, MA: Harvard University Press.

MEMO. 2013. "MEMO Project: Mass Extinction Monitoring Observatory." www.memoproject.org/. Accessed 5 December.

Moten, Fred. 2002. "Black Mo'nin.'" In *Loss: The Politics of Mourning*, edited by David Eng and David Kazanjian, 59–76. Berkeley: University of California Press.

Nixon, Rob. 2011. *Slow Violence and the Environmentalism of the Poor.* Cambridge, MA: Harvard University Press.

Plastow, Michael. 2012. "Retail Therapy: The Enjoyment of the Consumer." *British Journal of Psychotherapy* 28, no. 2: 204–20.

Radden, Jennifer, ed. 2002. *The Nature of Melancholy: From Aristotle to Kristeva*. New York: Oxford University Press.

Ramazani, Jahan. 1994. *Poetry of Mourning: The Modern Elegy from Hardy to Heaney*. Chicago: University of Chicago Press.

Riaño-Alcalá, Pilar. 2006. *Dwellers of Memory: Youth and Violence in Medellín, Colombia*. New Brunswick, NJ: Transaction Publishers.

Ricoeur, Paul. 1999. "Memory and Forgetting." In *Questioning Ethics: Contemporary Debates in Philosophy,* edited by Richard Keaney and Mark Dooley, 5–11. London: Routledge.

Samuels, David, Louise Meintjes, Ana Maria Ochoa, and Thomas Porcello. 2010. "Soundscapes: Towards a Sounded Anthropology." *Annual Review of Anthropology* 39: 329–45.

Sandberg, L. Anders, and Tor Sandberd, eds. 2010. *Climate Change – Who's Carrying the Burden? The Chilly Climates of the Global Environmental Dilemma.* Ottawa: Canadian Centre for Policy Alternatives.

Schafer, R. Murray. 1977. *The Tuning of the World: The Soundscape.* Toronto: McClelland and Stewart.

Smith, Mick. 2001. "Environmental Anamnesis: Walter Benjamin and the Ethics of Extinction." *Environmental Ethics* 23, no. 4: 359–76.

Spargo, R. Clifton. 2004. *The Ethics of Mourning: Grief and Responsibility in Elegiac Literature*. Baltimore: Johns Hopkins University Press.

Stanescu, James. 2012. "Species Trouble: Judith Butler, Mourning, and the Precarious Lives of Animals." *Hypatia* 27, no. 3: 567–82.

Tracey, Andrew. 1970. *How to Play the Mbira (dza vadzimu).* Roodepoort, Transvaal: International Library of African Music.

Windle, Phyllis. 1992. "The Ecology of Grief." *Bio Science* 42, no. 5: 363–6.

# 10 Emotional Solidarity

## *Ecological Emotional Outlaws Mourning Environmental Loss and Empowering Positive Change*

LISA KRETZ

## Introduction: Justified Mourning

The ecological crisis provides no shortage of evidence for justified mourning. For a non-exhaustive list, consider the following: overpopulation (more than 200,000 people added every day); global warming (global ice cap melting, sea level rise, increasing catastrophic natural disasters); deforestation (32 million acres annually); unsustainable agriculture[1] (the abominable treatment of non-human animals aside, current farming practices are responsible for 70 per cent of the pollution of United States rivers and streams); unsustainable transportation (a single car emits 12,000 pounds of carbon dioxide every year in the form of exhaust; in the United States cars emit roughly the same amount of carbon dioxide as coal-burning power plants); human-caused environmental accidents (e.g. Exxon Valdez and the BP Oil disaster); coal mining (mountain top removal and strip mining); invasive species (400 of the endangered 958 species listed in the United States are at risk because of competition with alien species); overfishing (90 per cent of the ocean's large fish have been fished out of their natural habitats); and damming waterways (reservoir-induced seismicity refers to dams causing earthquakes; the Three Gorges reservoir is built over two major fault lines – hundreds of small tremors have occurred since it opened) (Schwarzfeld 2013).

The Fifth Assessment Report of the Intergovernmental Panel on Climate Change attests to widespread scientific agreement that humans are, indeed, responsible for dangerous levels of global climate change (IPCC 2013). Given that ecological harms of unprecedented proportions continue daily, one can argue that an adequate response to the global climate change crisis has not yet happened. In terms of escalating oil consumption alone, consider that over the course of a single lifetime humans have burned 97 per cent of all the oil that has ever been burned (Thompson 2010). The ecological ramifications of overpopulation coupled with energy addiction are manifesting themselves in increasing extreme weather events, which in turn are evident in the increasing numbers of environmental refugees. The American Association for the Advancement of Science warns that the United Nations predicts that by 2020 we will have 50 million environmental refugees (Zelman 2011).

Humans are driving extinction faster than species can evolve; for the first time since the disappearance of the dinosaurs the rate of species extinction is faster than the rate of new species evolving (Jowit 2010). This rate of extinction of species is driven by destruction of natural habitats, hunting, the spread of alien predators and disease, and climate change (ibid.). Insufficient progress has been made toward sustainable solutions in spite of the fact that anthropogenically caused ecological harms are destroying the very ecological systems that underwrite the possibility for life as we recognize and value it (Fiala 2010). Val Plumwood (2006) argues that the social change necessary for reducing harmful human impacts, at minimum to a level that facilitates the future survival of human society, has not occurred. The emotional impact of such losses, coupled with the alarming lack of action to remedy the situation, can lead directly to despair and avoidance if there is an absence of community support and action. Relatedly, environmental educator Elizabeth Andre (2011) asks in her work on ecological mourning: "Could it be that people won't deal with the climate issue until they have a way to deal with the emotions that come along with it?" (2).

Part of my motivation in what follows is to assess the emotional capacities that might limit action if they fail to be cultivated and activated in fruitful ways. The types of emotional community support I envision are diverse, interact in multifarious ways, and occur at multiple levels simultaneously. My account is influenced by Susan Sherwin's (2008) conceptualization of public ethics. Sherwin argues that the ethics required to deal with the many global threats to human well-being, including environmental threats, require looking "beyond the moral responsibilities of individuals and consider[ing] the ways in which social organizations of all sorts (including community groups, corporations, governments, and international bodies) also must alter their behaviors" (9). There are communities at local, national, international, and global levels that are politically, socially, ecologically, spiritually, and emotionally organized in manifold and overlapping ways. Emotional dimensions of life simultaneously shape and are shaped by, reflect and are reflected by all of these levels. Although I make a case for seriously considering the role of emotion as it relates to ecological harm and political empowerment – or lack thereof – my account is both exploratory and truncated. It is by no means meant to be definitive – rather, I envision a contributory perspective open to perpetual revision.

Environmental mourning and the constellations of emotion attendant to such experiences of loss are understudied in Western, analytic, philosophical writing. This volume creates the opportunity for academic engagement with the emotional fallout that comes from witnessing the violent death of the inhabitants and living systems of planet Earth. I wish to make the case that giving voice to this loss, and the sadness that accompanies it, is essential for emotional honesty (wherein emotions are not repressed, denied, or ignored) and morally adequate responses. Community mourning can serve to provide support for, and galvanize the strength of, those grappling with ecological loss. Importantly, emotional solidarity through community support can validate "outlaw" emotions. Community support can likewise counter the threat of ecological despair and terror management, where management techniques often involve denial and refusal to deal with the source of the terror. The form of ecological mourning I address in what follows is symptomatic of harmful human-ecological relationships; movement toward long-term healing involves community empowerment for positive, protective, and proactive ecological action. Communities can be as small as two and as large as the global community.

People who mourn beloved animals, environments, rivers, mountains, trees, insects, as well as entire species, involve circumstances wherein rebirth of *this* particular individual or *this* particular species is not possible (barring appeals to various religious traditions). But one cannot stop here, for in the case of the living planet and those inhabitants and systems that people value, there are many complex and exquisite forms of life that are not yet dead. Renewal remains possible for those not yet destroyed, for those who could help weave together future planetary health. Backward-looking mourning at what once was simply cannot be the place where we stop. Possibilities for renewal demand forward-looking orientations where emotions serve to empower people.[2] I will explore the justified emotion of anger in response to the ecological injustices that ecological mourning is predicated upon. Additionally, I identify an emotional orientation toward hope as a method for fighting ecological despair and terror. Anger and hope are not postulated as being sufficient for positive change. Rather, their potential as justified responses that can help fuel positive change is explored.

I begin with a political analysis of emotion. A more nuanced account of emotion offers an explanation of the importance of emotional solidarity as a political tool for effecting positive change. Attending to the historical development of how to conceptualize emotion in the Western tradition – the wrongful relegation of emotion to the domain of the irrational and the powerful, pervasive, subtle but explicit politics of emotion – helps provide a rationale as to why there has not been a groundswell of public articulations of ecological mourning.[3] At the heart of my analysis is a move toward political recognition of the epistemic import of emotional responses, nurturing appropriate emotional responses to the suffering of human and non-human others, and orienting ourselves emotionally for facilitating positive, ethically grounded change.[4]

## Accounting for Emotion: Background Context

What emotion amounts to has a rich and complex history. Methods for sense-making regarding what precisely emotions are vary significantly, ranging from definition by scientific reduction to isolated events, to identifying constituent parts in synergistic interaction, to identifying wholes that are not reducible to their parts, to identifying experiences so complex they are potentially better gestured at by art than pinned down piecemeal by science (Brennan 2004). The range of experience meant to be captured likewise shifts when experiences of emotion are explored within the restraining contexts of individual human domains, group human levels, varied individual and group ecological levels, or any combination therein.[5] Understandings of emotion also can vary depending on the intended applications of said understanding; for a non-exhaustive list consider: psychology, philosophy, politics, geography, education, psychiatry, sociology, and literature. Differing characterizations among theorists demand an overarching conceptualization of emotion that is dynamic, pluralistic, context sensitive, and occasionally frustratingly elusive. Moreover there are considerations of relationships (if distinctions are accepted at all) between affects, passions, feelings, and emotions. I use emotion as a broad umbrella term with the intent of capturing much of this pluralism.

I think of the varying understandings of emotion as each being a contribution to a nuanced story of emotion. Each, exempt from the others, is insufficient to the task of highlighting the varied meanings and understandings that identify facets of a fluid and evolving concept. I suggest that defining emotion with a precision that captures a consistent meaning over expanses of times and domains would invite a static understanding of emotion. Such an understanding fails to reflect the diversity of reference this concept generates. Rather than offering such a definition, the elements of emotion I take to be relevant for development of my position will be articulated as my argument unfolds and will include cognitive, psychological, social, and political dimensions. Appealing to varied understandings of emotion can be problematic when those concepts contain contradictory elements. However, the multiple understandings of emotion that I appeal to in my position do not adopt understandings of emotion that necessarily conflict. Or, stated differently, on a generous reading of the conceptualization of emotion I offer, internal conflict need not, of necessity, be present. Implicit in my account is a conceptualization of emotion wherein there are moral dimensions to emotional experience. I am focusing on emotions that a) are responses to morally charged events, and b) can be reflected on and assessed with regard to their justifiability.

In the history of Western analytic philosophy, we see a recurring value dualism between reason and emotion. Emotion is often taken to belong to the domain of the irrational and is contrasted with the exercise of reason. Insofar as this dualism is gendered, women are taken, predominantly, to be identified with the emotional dimensions of human experience. Men, in contrast, are predominantly identified with the rational dimensions of human experience.[6] Given the historical pervasiveness of gender oppression and attendant devaluing of that which is associated with expressions of feminine gender, there continues to be a pronounced denigration of the epistemic worth of emotions.[7] Systematic emotional oppression happens when a group's justified emotional response is denied uptake due to group membership. Historically, people have faced this due to gender and ethnicity. For example, historically women were grouped by their gender and oppressed as a group. I wish to add to the analysis of varied forms of oppression a concern with emotional oppression in the context of ecological harms. People who wish to give voice to the harms to the Earth through mourning are often emotionally oppressed. The systematic denial of this attempt at emotional expression is disempowering. Part of what is at work, I contend, is a devaluing of emotion generally.

Alison Jaggar (1997) points out that within Western philosophy emotions are more often than not considered subversive to knowledge. Elizabeth Spelman (1989) also notes that there is considerable anxiety about emotions in Western philosophy due to a tendency to see them as interfering with the successful function of reason. James Jasper (1998) contends that "there is still a taint or suspicion of irrationality surrounding most emotions" (408).

I use this analysis as a lens to see what it may contribute to understanding why there has been so little attention to the emotional experience of mourning at witnessing the death of existing life forms and systems on this planet. Ashlee Cunsolo queries, in this collection and elsewhere, why the grief and mourning experienced in response to the anthropogenic destruction of Earth's inhabitants and living systems appear strangely silenced in public discourse. I suggest the undermining, devaluing, and political dismissal of emotion plays a role in the rarity of successful articulations of group emotion as a platform for social change in the Western tradition. Such analysis also helps identify the sometimes subtle, but no less deeply disempowering, lack of uptake that occurs when such attempts at articulating emotion do manage to arise.

## Reasonable Emotions and Emotional Reasons

I contend that most "emotion versus reason" disputes are unnecessary, as they hang on a misunderstanding that can be clarified by a more nuanced awareness of the relationships between reason and emotion. In *Anger and Insubordination*, Spelman (1989) quotes Aristotle: anyone "who does not get angry when there is reason to be angry, or does not get angry in the right way at the right time and with the right people, is a dolt" (263). If I stubbed my toe painfully on a rock while walking, and started yelling angrily at the next person who crossed my path – thereby holding them blameworthy for the event and the rightful recipient of my angry diatribe – then I would be acting foolishly (and unkindly). The reason why such action is recognized as foolish is because emotions have a relation to circumstances wherein they can be either appropriate or inappropriate. On the "dumb" view of emotions, emotions are simply "dumb" occurrences. Emotions are taken to be "like feelings of dizziness or spasms of pain since they do not involve any kind of cognitive state. According to this view emotions are, quite literally, 'dumb' events" (ibid., 265). On the "dumb" view, Aristotle's reflection fails to make sense. In contrast, cognitivist views of the emotions – which have seen an intense revival – insist judgments, beliefs, or some kind of cognitive state, are constituent of emotions. When people are happy, sad, jealous, or angry it is typically at or about something or someone. Additionally these emotions are regarded as "appropriate or inappropriate, reasonable or unreasonable, justified or unjustified, by others or by oneself" (ibid.). Given that the "dumb" view of emotions is incapable of accounting for the aboutness of emotions, and the ways in which emotions are evaluated and judged, it is incorrect and should be abandoned. Insofar as a) our emotional lives require narratives that explain why our feelings are justified and when they should be acted on, b) reasoning provides such explanations and justifications and c) without the desire that accompanies emotion there would be no will to act, we must recognize the cognitive dimensions of emotions, and the emotional dimensions of cognitions.

Recognition of the relevance of emotion to our moral lives is not without philosophical precedence in the Western tradition. For example, seventeenth-century Scottish philosopher David Hume highlighted the necessary role of emotion in our ethical lives, arguing that no action is possible without an emotional impetus – without a desire for one thing over another we would be perpetually indifferent to states of affairs (Hume 2002). Likewise, Jaggar notes, "values presuppose emotions to the extent that emotions provide the experiential basis for values. If we had no emotional responses to the world, it is inconceivable that we should ever come to value one state of affairs more highly than another" (391). Moreover, given cognitive, communicative dimensions of emotional expression, no emotion can successfully be communicated in a manner understood as justified without it being a reasonable response to a state of the world. Jaggar highlights a mutually constituting relation between reason and emotion through her demand that we attend to the social construction of emotion. Attributions of emotions only happen in the context of shared linguistic resources with associated concepts; children are taught what counts as appropriate expressions of emotion with the intent of generating the habituation of appropriate identification and expression of a culturally identified set of emotional resources. Emotions are constituted, in part, by shared social meanings, norms, and expectations regarding appropriate behaviour and response (Averill 1980 in Jasper 1998). Such a stance is required to account for the considerable variability historically and cross-culturally in the situational causes, experience, meaning, display, and regulation of emotion (Thoits 1989).[8] Any viable view of human experience, I contend, must minimally recognize the necessity of an intimately interwoven symbiotic relation between emotion and reason wherein both provide epistemic insight.

Furthermore, an approach seeking to facilitate action must of necessity attend to emotion. A dominant, but nonetheless erroneous, assumption is that once individuals become sufficiently knowledgeable about their contribution to an environmental problem they will shift their attitude and then shift their behaviour to reflect that new attitude toward the problem. In other words, knowledge will lead to a change in attitude, and a change in attitude will lead to a change in behaviour. The "knowledge-attitude-behaviour" method has been widely critiqued given that increased knowledge about nature does not, of necessity, lead to either a) a favourable attitude toward nature or b) action on behalf of nature (Goralnik and Nelson 2011).[9] Cognitive agreement, on its own, does not result in action – emotional engagement is required (Jasper 1998; Kals et al. 1999). Lissy Goralnik and Michael Nelson highlight the crucial role of care and a sense of community in environmental theories meant to generate action, and recommend focusing on relationships that inspire care and empathy. An approach to morality where reason and emotion play a necessary role is indicated by contemporary brain research (Goralnik and Nelson 2011). Chrisoula Andreou (2007) notes that the "prevailing view among empirically oriented moral philosophers is that morality is grounded in sentiment ... at least for human beings, the capacity to make genuine moral judgments depends on related emotional capacities" (47).[10] Emotional intelligence is thus crucial for motivating, generating, and responding to and with emotions that empower. I borrow the term emotional intelligence from Karen Warren (2005) who argues "Moral emotions are part of what psychologists have called 'emotional intelligence' ... [and are] essential to moral reasoning" (271). Emotional intelligence is important for ethics, ethical reasoning, and ethical decision-making; ethical motivation, reasoning, and practice require that rational and emotional intelligence operate in concert (Warren 2005). Warren is here utilizing the work of Daniel Goleman (2005) who identifies limitations to IQ tests – which focus solely on rationality – and envisions emotional intelligence as a complementary capacity. To facilitate fecund imaginaries it is far better, I suggest, to explore emotional intelligences – in order to capture the pluralistic and diverse capacities of emotional understandings, experiences, and knowings.[11] The existing emotional shaping of our lives and the potential for intentional emotional self-configuration are, as of yet, underutilized and extremely powerful sources for motivating positive environmental action.

## The Politics of Emotion

The successful expression of some types of emotion is a more fragile political achievement than many recognize (Campbell 1994, 47). Sue Campbell's (1994, 1997) account of emotional expression makes explicit the political dimensions inherent in public attempts to articulate our emotional lives. She does this by calling attention to the role of the would-be expresser(s) of an emotion, the role of the would-be interpreter(s) of an emotion, relevant context, and the attendant implicit power dynamics. The success of emotional expression hangs not just in a personal attempt at articulation but also in whether it is recognized as a successful expression of that emotion, whether it is recognized as a reasonable assessment, whether it is given "uptake."

Which emotions are socially recognized and validated, which emotions are given "uptake," is importantly politically grounded. In *A Note on Anger*, Frye (1983) introduces the notion of "uptake" with respect to the attempted expression of anger. She highlights that two parties are involved when it comes to successful expression, both the person attempting to express anger, for example, and the person who either recognizes it as anger or does not. In the example Frye gives, a woman attempts to express her anger and the response she receives is a character assault – she is called crazy. Frye notes that the anger and its claims are not met; rather, the topic was changed from what the woman takes to be an expression of her anger to the interpreter's reading of it as evidence of questionable sanity. Deprived of uptake her anger "is left as just a burst of expression of individual feeling. As a social act, an act of communication, it just doesn't happen" (89). As such, we require a theory that addresses "how resources for securing uptake can be unequally distributed so as to reinforce existing patterns of oppression, and how particular emotive criticisms can also serve this political goal" (Campbell 1994, 54). This is particularly important given that emotional work is often taken to be primarily women's work (Jasper 1998). It is a form of work both expected and undervalued; that is, if it is valued at all – for a tendency to perform the work of navigating emotional life is often written off as a tendency to focus on irrational responses. In contrast to this view, emotional work can be recognized as involving an inherently difficult skill to hone, a skill that requires attending to the complex emotional language of interaction central to healthy relationships and identities. To adequately defend emotions it is essential to understand that the "means by which we express our psychological states" – the expressive resources themselves – can be denied or used for dismissal (Campbell 1997, 48).

As a case in point, dominant culture in North America encourages a subset of women (those who fit the criteria for the Kantian feminine) to blush, cry, and smile – and these are promoted as feminine attributes (Campbell 1994, 57). Campbell identifies this set of expressive resources as being reflective of "'the Kantian feminine,' an ideal of a woman formed by race and upper class privilege and applicable mainly to such women" (ibid., 56). How emotional resources, both expressive and interpretive, are complicated through considerations of intersectional oppression needs further analysis. Moreover, although the expressive and interpretive constraints that tend to be attached to masculine and transgender emotive repertoires are beyond the scope of this chapter, they are certainly important areas for additional investigation. Campbell's example is illustrative of one way in which dismissal can work, and helpfully identifies some of the power dynamics at work in attempts at emotional expression.

Insofar as women are encouraged to blush, cry, and smile, alternative expressive resources may not be encouraged or recognized as meaningful when attempts are made to exercise them. When women employ these "feminine" expressive resources they can be taken as symptomatic of sentimentalism and over-emotionality, and thus grounds for dismissal (Campbell 1994). The duplicitousness here is marked. Women are expected to be emotional to satisfy gender requirements for ideal expression and intelligibility, but those very same expressive capacities are used as a ground for dismissal, hence the significance of a focus on the potential authority of our anger, of others to interpret us, and of our own judgments and experience (ibid.). Another example of a group whose expressive resources are taken to be "justification" for dismissal is the caricature of bleeding-heart environmentalists. Think here of an environmentalist attempting to express ecological mourning to an unreceptive individual, group, or even government. They might seek to express this emotion through attempting to articulate their suffocating frustration and the press of despair. Uptake denial might take the form of the claim that "You are being overly dramatic. The world is continuing on just fine. You should stop being so negative." Here deep concerns are dismissed through the accusation of over-emoting. Moreover, an environmentalist group's attempt at expression is further delegitimized by the charge of insincerity that the claim of being "overly dramatic" carries. For the successful expression of emotion one needs uptake, and uptake requires a collaboration of interpreters (ibid.).

Given the importance of uptake for successful emotion formation and articulation it becomes crucial to establish communities of support. In particular, I am thinking of supportive communities that will recognize mourning as a fitting response to the current ecological massacre, and will provide uptake for legitimate attempts at emotional expression.[12] This is not to say all people must of necessity mourn but, rather, that those who do mourn have justification for doing so and deserve uptake. Emotional expression involves attempting to identify and communicate what is salient, what is taken to be significant; as such, dismissal can serve as a powerful political tool meant to persuade already oppressed groups and individuals that their viewpoint is dismissible (ibid.). Campbell highlights the experience of dismissal – being dismissed is when "what we do or say, as assessed by what we would have described as our intentions in that situation, is either not taken seriously or not regarded at all in the context in which it is meant to have its effect ... Put more simply, if no one takes my anger seriously by making any attempt to account for his or her behavior or to change it, but, instead, characterizes me as upset and oversensitive, I may be unsure, in retrospect, of how to best describe my behavior" (ibid., 49). When power differentials work against the emotional uptake of members of certain groups we are faced with systematic epistemic denial. The epistemic denial is denial of what is taken to be important to one's own experience. It is to deny the legitimacy of what has been marked as salient in one's own life. Moreover, those who inhabit positions of social and economic dominance may seek to defuse the motivational power of justified emotional responses that demand recognition of the experience of non-dominant groups. That oppressed groups are disproportionately subject to the systematic denial of uptake of emotions that empower – such as anger – is well documented (Campbell 1994, 1997; Spelman 1989; Lorde 1984). Thus we can see how it comes to be that reasonable emotional responses to the current large-scale, human destruction of the environment (such as mourning) are politically dismissed. There is no global consensus on necessary revisions to current fundamentally unsustainable climate change policies (or the lack thereof), which are premised upon likewise unsustainable social, political, and economic systems. Lack of leadership and foresight, and the blatant failure to show a basic concern for future generations, can reasonably lead to a sense of mourning for all that has already been lost, at least for those who are honestly engaging emotionally with the breadth and depth of ecological destruction at human hands.

Jaggar (1997) argues that people who are marginalized are often better placed to think critically about dominant paradigms than those in dominant positions. (Also see Lorde, 1984.) Those who experience conventionally unacceptable or what she calls "outlaw emotions" are often "subordinated individuals who pay a disproportionately high price for maintaining the status quo. The social situation of such people makes them unable to experience the conventionally prescribed emotions" (Jagger 1997, 396). Although the examples I am about to articulate are importantly different from racism and sexism, they are nevertheless circumstances that potentially merit the critique of emotional outlaws. Consider the following: an inability to take pleasure in the consumeristic exchange of unnecessary goods destined for the dump at holiday time, or responding to a colleague's "winter getaway" plans by questioning the morality of *vacations* that involve flights generating obscene greenhouse gas emissions, or sadness at the woefully unreflective proliferation of procreation given an already unsustainable population,[13] or articulations of mourning for those creatures, spaces, and processes devastated by the anthropogenic destruction of nature. Granted, the reasons for any of these activities are more nuanced than a general analysis can give, and the particulars of the cases will need attending to. Nonetheless, the examples can still serve as potential points of departure for exploring ecologically rooted outlaw emotions. Outlaw emotions can be epistemically fruitful in that they help develop a perspective that is critical of dominant paradigms; ways of seeing that have been lifted to the status of "facts" are shown to be constructions – constructions that are to the benefit of some and the detriment of others (Jaggar 1997). Such considerations are especially important given that emotions are managed to fit social expectations, which can keep in place particular social, political, and economic conditions and agendas (Norgaard 2006, 384). Emotions are both a primary site of social control and political action that can be utilized to mobilize resistance and liberation (Boler 1999, xv). The transformative power of emotions is largely untapped in Western cultures that encourage fear and control of emotions.

Kari Marie Norgaard (2006) explores the social organization of climate change denial, and sees it as being kept in place through emotional norms shaping what is perceived as socially acceptable to feel. Denial of climate change is socially required in many environments, given the gravity of the harms climate change brings and the cloud it casts on what would otherwise be light and un-troubling, and therefore socially permissible, conversation. Outlaw emotions, when articulated and validated, can serve to call into question and alter current problematic anti-ecological emotion norms. Mourning and a desire to protect ecological others are morally appropriate responses to the ecological crisis. Thus, they should function as the new socially condoned and validated emotional norms as opposed to anomalies. Creating such norms can take place, at present, through the efforts of ecological emotional outlaws. The concept of the emotional outlaw is further elucidated through the work of Sara Ahmed. Ahmed helpfully flags the political operations of social communities and attendant emotion mores. She highlights the ways in which a failure to share a dominant emotional response is a failure to align with a community; as such it generates alienation (2008, 10–11). For example, chastizing the sexist "joketeller" in a roomful of sexists renders one a feminist killjoy (Jaggar 1997, 396; Ahmed 2008, 13). Feminist killjoys provide subversive observations that undermine dominant conceptions of the status quo; they illuminate ways in which social norms are constructed so as to obscure the experience of subordinated people (Jaggar 1997, 397). People seeking to mourn what has been lost are currently emotional outlaws to the extent that they are systematically denied uptake and resultantly disempowered. The possibilities for outlaw emotions extend beyond mourning, but given the focus of the volume I will focus on mourning and directly related emotions.

Andre (2011) is an environmental educator. In her presentations on a) why despair is so common in environmental circles and b) how to combat that despair, she senses that "people are eager to shine some light into the dark corners of their hearts and minds where they've been hiding sorrow, guilt, and feelings of hopelessness. In most social situations, talking about these dark topics would be outside of social norms ... Creating a forum where the stated purpose was to examine these feelings freed participants from these restrictive norms" (83). Andre previously had to steel herself against the negative responses to her worries and sadness about human-generated ecological decimation, but found that "sharing this mutual experience with others helps me to not feel so alone there" (84). Through the act of emotional solidarity a counter to isolated mourning was discovered, and a new source of strength and acceptance was revealed.

At a time when there has yet to be a global ethical evolution of compassionate consciousness, those who mourn the massive destruction of the ecosystems and species that populated planet Earth are emotional outlaws. Their insights are epistemically crucial in that they are honestly engaging with emotions around the increasing death and destruction of the planet and its inhabitants. This involves emotional integrity, bravery, and openness to responding to even the hard truths – the ones that justify mourning, threaten with despair, and make demands that we share our environmental grief and mourning with others (Cunsolo chapter 7). I am concerned about the silencing that results from lack of uptake. Patricia Monture-Okanee (1992), a Mohawk woman, writes of her experience of narratives reflecting only the views of colonizers as a form of silencing: "The continual denial of our experience at every corner, at every turn, from education at residential schools through to universities, is violence. The denial of my experience batters me from all directions. Because others have the power to define my existence, experience, and even my feelings, I am left with no place to stand and validly construct my reality. That is the violence of silence" (197–8). The social construction of emotions makes clear the importance of validating the experience of those who are unjustly silenced. To counter inaction we must strategize against the deflationary silencing of epistemically grounded, and ecologically sensitive emotions, especially given that environmental emotional outlaws are privy to crucial insights. The public articulation of outlaw emotions is a political act; it is a demand for recognition.

One method for identifying such concerns is through forming ecological mourning groups to give space for voicing and validating experience that is not yet widely validated. The function of these groups could be somewhat analogous to women's critical consciousness-raising groups that provide space for voicing and validating women's experience. Oppressed groups generally benefit from safe spaces for voicing and validating non-dominant experiences. The assimilationist expectations of dominant experience serve to silence, perpetuate epistemic error about experience (because experience is not monolithic), and repress diversity (Young 2009). Consciousness-raising groups not only validated the experiences of women, they helped provide a ground for political action addressing issues brought to the fore by second-wave feminism. Prior to the activity of consciousness-raising groups there was not a language to establish particular harms (e.g. intersectional oppression, women's double day, emotional labour) and there wasn't an understanding of the complex ways that individual harms were actually points in an interwoven systematic structure designed to repress women. Supportive community helped give shape, in language, to what could not be articulated previously. To stop a harm, you first have to be able to identify it. In this way emotional outlaws can band together to give name to what is impossible to identify without the systematic analysis provided through group insight; both personal validation of grief and public articulations of mourning can be fruitfully articulated. The means by which they are articulated are only as limited as our imaginations: protest marches, performances, sit-ins, petitions, books/articles, letters to the editor, documentaries, visual and auditory art, not-for-profit organizations, coalitions, capacity building, exploration and validation of different experiences and expressions of grief given different communities, ethnicities, geographies and so on.

An effective example of mourning motivating action is the Mothers of the Disappeared movement. It is telling of the political silencing of emotion in that the legitimate claims and demands of the Argentinian women were initially met with a character assault that called into question their sanity (much like in Frye's example). Government officials tried to marginalize and trivialize the efforts of the mothers searching for their children by calling them "las locas" which is translated as "the madwomen" (Kurtz 2010). Fortunately those efforts at silencing were not successful.

My analysis of emotion helps explain why large-scale, socially validated expressions of mourning for environmental loss have not been forthcoming despite clear indications that such mourning is in fact appropriate. It may be tied, at least in dominant Western paradigms to a) a de-valuing of the validity of emotions and b) a political denial of the attempt to express outlaw emotions. Public articulations of outlaw ecological emotions such as mourning simultaneously reflect the worth of nature and demand accountability.

## Emotional Solidarity: What to Do with Mourning

Mick Smith (2001) contends mourning is a painful exercise in memory, where the pain occurs because the other is no longer immediately present to us (Paul Ricoeur 1999 in Smith 2001). Mourning is taken to be an ethical relation, a reconciliation with the past and its loss, a coming to terms with death and extinction (Smith 2001). Mourning is contrasted with melancholy, which, following Sigmund Freud, is taken to be a pathological condition which occurs when we interiorize that which has been lost. "The melancholic ego wants to hold onto the past by consuming and fixing it in themselves, by becoming one with it. But the inevitable result of this impossible desire is that the subject becomes caught up in a depressive and repetitive spiral." Smith highlights that environmentalism "exemplifies a loss of faith in the ideology of 'progress,' which everywhere runs up against the reality of contemporary life, of road congestion and radiation, genetic modification and cancer producing chemicals, poisonous emissions and extinctions" (375). The danger is that, broken-hearted about ecological decimation, "we withdraw into ourselves, reciting the litany of those lost in our abject environmental melancholia" (ibid.). Rather than adopting a melancholic reflection on what is perceived as inevitable destruction, Smith recommends we take the fullness of memory to inspire political action against environmental harm. Building communities of support around shared experiences of mourning facilitates validation and capacity building for direct action against forces of environmental abuse. One such force of abuse is anthropocentric contributions to climate change.

As noted at the outset, Andre (2011) asks the tantalizing question "could it be that people won't deal with the climate issue until they have a way to deal with the emotions that come along with it?" (2). Climate denial may be a coping mechanism for those who don't want to be depressed and frightened by what we have done – and what we continue to do – to the planet and its members. It may serve as a temporary evasion for those wishing to push the margins of denial to the breaking point with the goal of avoiding despair, sorrow, anger, frustration, and accountability. Andre (2011) dares to ask "how do we deal with the grief of witnessing the destruction of places we love?" (30). Using autoethnography as her lens for exploration she articulates the following position: if the fate of numerous species is extinction, if they are in fact the walking dead, "their fate is already sealed, they just haven't quite disappeared yet. So mourn them and then move on. If there's truly nothing you can do to save them ... mourn their loss and then change your horizon. You can't be stuck in this state of continual grief ... all you're going to see is loss after loss after loss and there will be nothing you can do to help" (ibid., 73). Mourn then, what cannot be saved, but defend what remains and can be.

Mourning is a legitimate emotional response to large-scale, escalating, ecological devastation. Individuals sharing appropriate responses must identify communities of others who share in this understanding. Communities can help individuals support and validate each other; we must cradle the falling, and empower the rising, of each other. Aldo Leopold (1993) contends that "one of the penalties of an ecological education is that one lives alone in a world of wounds" (165). Emotional solidarity affords the opportunity to recognize we are not alone in our mourning, nor are we alone in our anger over the loss and the palpable pressing need for immediate action against such destructive forces. Not only do symptomatic wounds have to be tended to, simultaneously the source of those wounds must be stopped. Given that anger can be overwhelmed by grief, an analysis of anger is in order (Jasper 1998, 412). When justified anger at the source of loss is merited it can serve to mobilize and motivate.

## Justified Anger[14]

Effective protest is grounded in anger ... Anger nourishes hope and fuels rebellion, it presumes a judgment, presumes how things ought to be and aren't, presumes a caring. Emotion remains the best evidence of belief and value.

TURNER 1996, 21–2

Mary Midgley (1996) contends that for efforts in the direction of sustainability to be effective, they need to be supplemented by a direct, spontaneous moral feeling – a sense of outrage. Profound indignation is taken to be effective in environmental campaigning. Anger involves making a negative evaluative judgment where another is held responsible; their actions are taken to be blameworthy (Spelman 1989, 266). In imbalanced power relations, this judgment can be taken to be an act of insubordination.[15] Historically, uptake of anger was secured by privilege. Such considerations make clear that there is a politics of emotion. "If we recognize that judgments about wrong-doing are in some sense constitutive of anger, then we can begin to see that the censorship of anger is a way of short-circuiting, of censoring, judgments about wrong-doing" (272). Peter Lyman (1981) points out that anger is the "essential political emotion," thus silencing anger may repress political speech (Lyman in Spelman, ibid.). One of the most prevalent norms in social movements is the conviction that existing social conditions are unjust (Turner and Killian 1987, 242, in Jasper 1998). Anger is an appropriate response to experiencing or bearing witness to injustice. When others are oppressed, exploited, or otherwise treated unfairly they are justified in an angry response (Spelman 1989). So, too, are we justified in being angry about the oppression of non-human others.

Audre Lorde (1984) gives a politically motivating account of anger, arguing anger is an appropriate response to racism. Anger, I contend, is also an appropriate response to naturism. She highlights positive uses for anger as a spotlight for injustice that can be used in the service of change, as a site of learning, as loaded with information and energy, and as an existing forceful resource that can be used "against those oppressions, personal and institutional, which brought that anger into being" (124–7). She goes on to discuss the power of anger. "Focused with precision it can become a powerful source of energy serving progress and change. And when I speak of change, I do not mean a simple switch of positions or a temporary lessening of tensions ... I am speaking of a basic and radical alteration in those assumptions underlining our lives" (127). Lorde speaks of a paradigm shift.

Anger successfully conveyed is a political accomplishment. Hatred is to be distinguished from anger. "Hatred is the fury of those who do not share our goals, and its object is death and destruction. Anger is a grief of distortions between peers, and its object is change" (129). We hold others accountable when we are angry and we do so because, when the injustice is acknowledged, the need for change can then be defended. Using anger to fuel action is not to act out of blind rage. Successfully expressed anger is communication about an unjust present, and insofar as uptake is secured, it can prompt change for a just future. Anger need not be acted on immediately. Like a gas stove, all you need is for the pilot light to remain lit, at the ready for cranking up the heat when a boil is needed ... at the right time, in the right place, and directed at the right source. That source – in our case – is ecological injustice.

Fear about attempting to express anger is a legitimate worry. It can be rational, for those who are politically disempowered and attempting to articulate anger, to fear the response of those in positions of power (Spelman 1989). That being said, in activist communities it is important to realize that *the harder the pushback from those in positions of power the more indicative it is of the recognition that the activism being engaged in is a real threat to the status quo*. Politically tabling appropriate expressions of anger at injustice, and energizing related action with the motivational power of emotion, are strengths. These strengths can be developed and drawn upon so as to force those in positions of power to recognize the magnitude of our anger over an ever-quickening destruction of our planet. Anger can help provide a ground for mobilization (Jasper 1998, 409). Fear about the continuation of the status quo with regard to human-generated ecological harms is much more justified than fear of change. Allen Thompson (2010) makes the case that given the ecological threats we face there is a need for a new set of ecological virtues, which include courage and radical hope against despair and hopelessness. Emotional solidarity regarding anger toward ecological injustice awards moral authority to those concerned about the environment and carries the force of a socially-recognized and justified response. As the eminently brave Lorde articulated: "My fear of anger taught me nothing. Your fear of that anger will teach you nothing, also" (124).

## Justified Hope[16]

The backward-looking experience of justified mourning requires support to ground the forward-looking and empowering emotion of hope so as to help prevent slippage into despair. Despair, Paulo Freire (2011) notes, is inaction. Hope can function as a bridge from mourning to action. This bridge is particularly important given that mourning carries great sadness, which may lead to the inaction of despair. Emotional reserves, community support, and intentional emotional strategies for generating strength can help to resist the weight of this inaction. Rachel Carson serves as an exemplar of environmental virtue generally, and serves as an example of someone who successfully bridged mourning to action (Cafaro 2005). As Mitchell Thomashow (1996) highlights, Carlson was immersed in the "ecology of love and loss" in her study of the debilitating effects of DDT (dichlorodiphenyltrichloroethane) but utilized her positioning to become a role model of ecological citizenship and demonstrated the power of grassroots organizations to have a significant impact on public policy (41). The danger of despair is a legitimate concern though, particularly given that the enormity of global problems can lead to the belief that the ability of any one person to achieve change is hopeless (Courville and Piper 2004).

The threat of ecological terror and despair are pressing worries. Elin Kelsey (2012) notes that "despair about the planet has many labels, among them 'ecofatigue' and 'ecophobia'" (25). Some conservationists worry their messaging is counterproductive in that a continuing culture of hopelessness among conservation biologists will work against successfully mobilizing conservation action (Kelsey 2012). When pessimism prevails ecological successes fail to be highlighted and therefore fail to attract wide attention (ibid.). If this leads to despair the spiral of inaction is set in motion. Educators and conservation psychologists are finding that despair leads to terror management, where problems are downplayed, as in climate change denial, and hyper-materialism serves as an ineffective panacea (Kelsey 2013). A groundswell of positive stories of resilience that inspire ecological hope is needed to instigate positive action (ibid.). Strategic framing is essential, given that the way a situation is framed serves to constrain its set of perceived solutions (Kelsey 2003). Kelsey (2012) highlights that the continuing debates between those who contend "doom needs to be leavened by hope" and those "who maintain that an optimistic message plays down the crisis and justifies business as usual" are failing to acknowledge a crucial point (25). "What's at stake is more than what makes the best message; it's what makes the best conservation strategy. Chronicling demise offers little guidance. But if we tell stories about positive outcomes and share details of how they are achieved, the likelihood that they will be replicated will increase. Hope engenders conservation success, and success breeds more successes" (ibid.). The way a situation is framed, for example as a challenge as opposed to an impossibility, can mean the difference between inaction and positive action. In situations where one is facing slow uptake by a reluctant community of potential and necessary collaborators, the crucial role of generating communities of support to validate justified emotions is all the more evident.[17]

Andre (2011) discusses her own experience with ecological despair. In seeking out healthy methods for grappling with ecological destruction Andre recommends humility.[18] Humility enables a vision wherein we recognize our limits, and with those limits comes proposals for what is in fact possible and achievable – goals that can support and generate hope. She explains that, through the sharing of mutual experience, she is enabled to manage both the emotional and mental challenges of environmental education in an age of mass ecological devastation. Andre notes that through the process of becoming educated about the dimensions of anthropogenic ecological destruction, through delving deeply into the literature, she gained the vocabulary, theories, and models required to understand and articulate her own experience (83). "These tools allow me to enrich the discussions I have with friends, colleagues, and strangers at parties. I can sense that the people with whom I talk are excited to gain this vocabulary and framework; it seems to validate their experience, make them feel less alone, and give them the tools to reflect and share with others" (ibid.). Prior to addressing a problem, one needs tools to recognize the nature of the problem; sharing frameworks and vocabularies can contribute to this goal.

## Amplifying Action

The space between current facts and future possibilities must be broached with epistemic, emotional, and psychological care. Although justified, hope requires a factually grounded account of the current state of affairs – otherwise it is based on falsity; the associated emotional response can either engender despair or hope depending on how one is imagining future possibilities (Kretz 2013a). Thus, a clear sense of how the psychology of hope functions is essential for proactively generating emotional responses that enable rather than hinder positive environmental action.

When we are hopeful we act in hopeful ways, which makes it possible to achieve the hoped-for goal. Hope is recognized as being psychologically necessary for survival[19] as well as contagious; it is the sort of thing that we can pass on to each other.[20] Being a hopeful person characteristically results in: having many goals, being more successful at achieving your goals, experiencing less distress and more happiness, and having superior coping skills (Snyder 1995). Importantly, hope generates active coping, prevents disengagement from stressful situations and reduces denial, which is particularly important given the worry about climate change denial articulated above (Alloy, Abramson, and Chiara 2000 in Braithwaite 2004, 83). Insofar as we are interested in habituating hopeful attitudes and practices it is useful to note that studies show that hope can be taught (Cheavens et al. 2005, 126).

There is an essential connection between agency and hope (McGeer 2004, 103). In the absence of individual agency, a mourner cannot rationally imagine being able to work against the environmental harms that give cause for mourning. In the absence of agency at the group level, those who are mourning cannot hope to use that sadness as a platform for building capacity to work against environmental harm. Peter Drahos (2004) notes that hope leads to "a cycle of expectation, planning, and action that sees the agent explore the power of her agency" (22). It is a process that facilitates building upon, and enhancing, agency. For such agency to be successfully exercised, adequate peer scaffolding must be in place. Victoria McGeer (2004) explains peer scaffolding as the psychological reinforcement of other's effective agency through recognizing and respecting their hopes. She contends that support for others and for one's own hope is best accomplished through communities of mutually responsive, high-hope persons. There is particular moral need for empowering historically disenfranchised communities. Those in positions of unearned privilege can utilize their position of power to amplify the voices of the disenfranchised; through that very action work is done against the immoral hierarchy based on the myth of meritocracy (McIntosh 2000). Those who are historically disenfranchised have perspectival knowledge that is essential for working against harmful power imbalances. Conversations across difference are essential given that hope develops in relationship with others. Fulfillment of hope is directly dependent on wider circles of action by others (Drahos 2004, 20). I've argued elsewhere, drawing from Snyder, that: "Communities of good hope can be formed through supportive relations. Growing hope requires environments where people interact in a supportive atmosphere such that individual and collective goals can be met (Snyder 2005, 359). Through creating such environments people can increasingly perceive that they have both the agency and the pathways to succeed" (ibid.; Kretz 2013a). Conditions that ground hope's ability to grow include having a sense of coherence about your own life which is found in a supportive and understanding network of persons who provide care and a base for thinking positively, for finding new hope, and for identifying paths to its realization (Braithwaite 2004b, 133). In light of the ecological challenges currently being faced, and in light of the crucial role community plays for generating justified hope, I suggest that coming together to mourn ecological loss could simultaneously generate community capacity for responding to, and working against, ecological harm. Hopefulness is a psychological orientation that encourages success when it is used for energizing action in the direction of that which is hoped for.

Hopefulness and hopelessness function as self-fulfilling prophecies. In the absence of hope, no effort is made to achieve the hoped-for goal, which ensures the goal will not be achieved. If instead hopeful action is exercised, the possibility of the hoped-for goal can be realized, which gives further cause for hoping for further goals, and so on. Hope is a precondition for the tools for change to emerge and take hold.[21]

> If the idea of trying to move society to a sustainable condition seems too enormous an undertaking, many may conclude it is pointless to try and the resulting sense of hopeless malaise "turns the belief that society will not change into a self-fulfilling prophecy." In contrast, collective hope allows for more individuals to visualize goals, anticipate potential obstacles, and generate flexible plans for goal attainment. Therefore, social movements and groups create possibilities for change which could not have occurred without hope for the specific outcome (Milbrath 1995, 108 quoted in Lueck 2007, 253). By providing tools for plan creation, hope is a crucial part of the feedback loop that occurs between planning, action, and outcomes that generate or alter expectations and hope (Lueck 2007, 252). (Kretz 2013a)

With the growth of hopeful, active communities of support, positive action grows – action, which can then serve as fertile soil for ever more regeneration. To quote Sasha Courville and Nicola Piper, there is mounting evidence that "while neoliberal economic globalization gathers pace, so do dissenting social and political movements, which resist the current trend of marketization, privatization, liberalization, and deterioration of labor standards" (Gills and Galbraith 2000, in Courville and Piper 2004, 43).

An apathetic and passive response to ecological harms is encouraged by neoliberalism's rhetoric of market forces being the purveyors of meaning and value (Harvey 2007). Consumers are encouraged to adopt the ruse that happiness lies in consumption, while psychological evidence shows the relationship between consumption and personal happiness is a weak one (Durning 2009, 506–8). The two primary sources of human fulfillment are relationships and leisure, and opportunites for nurturing relationships and engaging in leisure time pursuits have dwindled under the pressure of market forces (ibid., 507–8). The attempt to satisfy social, psychological, and spiritual needs with material things is a fruitless endeavor (ibid.). The consume or decline argument – where human life is taken to require an economy like the current one, and a viable economy is taken to require the sorts of jobs currently available, and those jobs are taken to require consumption of ecological resources – is used as a forceful rhetorical tool to ensure business as usual (ibid., 509–13). The argument is flawed in so far as: a) Economic systems are human inventions and the current one is a theoretical model being tested. What that test has shown is that it is a model that undermines the welfare of the majority of Earth's living systems and members; b) Although we need exchange systems we do not need to use the current model; c) Jobs are necessary but the sorts of jobs currently available reflect an unsustainable relationship with the majority of Earth members (including a large number of oppressed humans). Alternative jobs that reflect care and respect can be created; and d) Jobs need not rest on the consumption of physical resources. It is necessary services that we should be most interested in having fulfilled and those services can be fulfilled in multiple ways that do not require the perpetual exploitative consumption of non-renewables (ibid.). Neoliberal ideology prompts a vapid form of pseudo-hope rooted largely in denial of the ecological ramifications of current practices.[22] Authentic hope, as juxtaposed with pseudo-hope premised on delusion and falsity, is necessarily rooted in world-reflective and world-mapping analysis of actual states of affairs. The actual state of affairs is that business as usual is resulting in the destruction of the living members and systems that underwrite life on this planet. If authentic hope – justified hope – is sought, then an alternative approach is needed.

I do not think that hope for the liberation of nature from an oppressive tyranny of economic growth premised on exploitation is misplaced (Kretz 2013a). Looking back at human history, we've seen movement after movement evolving in the direction of recognizing a greater diversity of those deserving of ethical consideration and respect; or as Leopold (1949) would say we evolve morally through expanding who we think of as fellow members of our ethical community. What seemed unimaginable one hundred years ago is commonplace in various locations today as efforts against racism, sexism, homophobia, ableism, classism, naturism and the like gain not only *de jure* but *de facto* hold as compassionate responses to others increase. "Institutions of hope move us collectively away from a social script that makes engagement in shaping our futures seem futile toward one in which we are expected to be active and responsible participants contributing to a vibrant civil society" (Braithwaite 2004a, 7). Hopeful vision roots radical change.

## Conclusion

Mourning is a strong emotional response to significant loss. Denial of the epistemic import of emotional insight is both rampant and ungrounded. I argue that emotion and rationality must necessarily act in tandem. Insofar as one accepts a cognitive account of the emotions, an argument can be developed that mourning is a justified, and appropriate, response to the destruction of the diverse ecosystems that support life as we currently know it. The fall into despair that mourning can prompt must be resisted. That resistance comes through emotional solidarity – building communities of support with those who honestly and emotionally engage with the violent assault of Mother Earth, and validate the legitimate sadness to which this loss should give rise. Acting as emotional outlaws, people can identify inappropriate emotional responses to the ecological crisis. Using community support, people can move toward embracing the energizing emotions of anger and hope.

We can demand political recognition and resolution of the injustice implicit in the destruction of life on Earth, and can strengthen the capacity for change through communities of support. Emotional solidarity must eventually, in building capacity, reach beyond those who are currently concerned about the environment. Critical mass requires, I contend, a shift in emotional consciousness. An emotional paradigm shift can happen in a heartbeat and good ideas can spread at the speed of sound. An environmental ethic energized by an intentional and robust engagement of emotional intelligence grounds future possibilities heretofore unimagined. Let's dare to feel with our whole selves so as to witness what engagement with the world in emotional and political fullness reveals.


## Acknowledgments

First, I offer my thanks to Ashlee Cunsolo and Karen Landman for their generous editorial support and encouragement. I also wish to thank Raymond Anthony for his close reading of, and insightful recommendations on, an early draft. I extend my gratitude to Joe Couture, James Kretz, David Kretz, Michelle Willms, Laura Mattiussi, Monique Dumont, Richard Ranger, and Jean-Francois Ranger for their constructive advice and emotional support while researching and writing. I am also indebted to the faculty and students at Michigan State University, with special thanks to Kyle Powys White and Michael O'Rourke; an invitation by the Philosophy Department to speak on this topic resulted in improved conceptual clarity and a more rigorous demand for making explicit practical applications. Finally I wish to thank Grand Valley State University and the University of Evansville for their institutional support while working on this project. In particular, my thanks are offered to Gail Vignola for her editorial assistance.


### NOTES

1 For an informative and extensive discussion of the negative impacts of agriculture as currently practised see the United Nations Report *Livestock's Long Shadow – Environmental Issues and Options* (2006) released by the Food and Agriculture Organization of the United Nations.

2 Here I am also alluding to forward-looking elements derived from that mourning – namely taking action in ways that work against the source of environmental harm. The temporal possibilities for mourning are varied. We can mourn that which is in the process of being destroyed as well as experience a sort of mourning in the making or anticipatory mourning wherein we mourn the likely losses on the horizon. I thank Ashlee Cunsolo for encouraging me to highlight these senses of mourning.
3 Or, perhaps more properly, why there has been failure to give adequate political uptake to attempts at communicating this loss thus far.
4 It is important to highlight, here at the outset, the limits of my analysis. My context is that of white middle-class female academic working in the global North. Both my experience and recommendations are informed and limited by this positioning. What I offer is one small contribution to a much larger, robustly inclusive dialogue that I hope occurs.
5 For geographically sensitive conceptualizations in particular see Davidson et al. 2005, Thrift 2004, and Cunsolo Willox et al. 2013.
6 For an extensive discussion of how the gendered reason/emotion dualism historically functioned in philosophy see Genevieve Lloyd's *Man of Reason* (1984). Lloyd's analysis includes the work of Plato, St Augustine, Thomas Aquinas, Francis Bacon, René Descartes, David Hume, Jean-Jacques Rousseau, Immanuel Kant, Georg Hegel, Jean-Paul Sartre, and Simone de Beauvoir.
7 I am here relying on Marilyn Frye's (1983) conceptualization of oppression where oppression involves a group that is harmed systematically due to their group membership while another group benefits; the harm is related to disempowerment and the advantage is related to having power over another (Frye 1983). For a nuanced analysis that problematizes privilege conceptualized solely as an advantage see Peggy McIntosh. She worries about the ways in which power over another often includes an implicit arrogance and ignorance, noting that the role of oppressor is not one anyone should see as desirable (McIntosh 2000).
8 This is not to say there are no basic universal emotions across the human species – there are; rather, it is to say that social construction and socialization play a key role in explaining the diverse meanings of myriad emotions (Thoits 1989).
9 For evidence of a disconnect between environmental knowledge/attitude and environmental behaviour that reflects this knowledge see: Bickman 1972; Costanzo et al. 1986; Finger 1994; Geller 1981; Geller et al. 1983; Hsu 2004; Hungerford and Volk 1990; Kollmuss and Agyeman 2002; McKenzie-Mohr 2000; and Sia, Hungerford, and Tomera 1985/86.

10 For a more in-depth discussion of above and related material see "Climate Change: Bridging the Theory-Action Gap" (Kretz 2012).
11 My thanks to Ashlee Cunsolo for suggesting improved conceptualization may be facilitated through articulating multiple emotional intelligences (plural) as opposed to emotional intelligence (singular).
12 Psychological care will need to be exercised with regard to appropriately navigating mourning, including the temporal dimensions of healing. Mourning is often a slow process; time plays a crucial role in healing.
13 Such considerations must, of course, reflect the differential distribution of consumption and the impacts of systemic oppression (Shiva 2005).
14 To clarify, I'm not saying we *should mourn* so as *to get* angry and hopeful – I'm not saying we always have to get sad before we get mad. One might be angry regarding massive, human-caused, ecological destruction without a strong experience of mourning prior to becoming angry. Rather, I make the case that mourning is a reasonable response to massive, human-caused, ecological destruction. Following this recognition, we ought to be proactive about how we want to reflect, tend to, implement and utilize this emotional knowledge. We must decide what to do with this knowledge and I'm recommending that we choose ways of engaging with this knowledge that empower. In the next section I am not advocating anger as a long-term goal or a stopping place. It is an intense emotion, and we need to funnel that energy to remedying the injustice that is the source of that anger. Anger is an emotion that, ideally, one renders no longer necessary by remedying the source of the anger.
15 A striking exception to the relegation of emotional expression to oppressed groups, says Spelman (1989), is anger; "their anger will not be tolerated: the possibility of their being angry will be excluded by the dominant group's profile of them" (264).
16 In the following two sections I cover some of the same theoretical territory I address in "Hope in Environmental Philosophy" but it is reframed in light of my argument in the present context (Kretz 2013a).
17 My thanks to Raymond Anthony for posing the question that this point answers.
18 Although humility, much like anger and hope, can be parsed as an environmental virtue, I am refraining from adopting a moral environmental virtue lens throughout my analysis because not all environmental virtues are emotions, and a full analysis of environmental virtues – including the role of emotion within such accounts – is beyond the scope of this chapter.
19 See for example: Stotland 1969; Farber in Snyder 1995; Freire 2011; Eliott 2005; McGeer 2004.
20 See for example: Cunningham 2004; Eliott 2005; V. Braithwaite 2004a; Snyder 1995.

21 I offer my thanks to Raymond Anthony for making this point explicit.
22 My thanks to a blind reviewer for recommending exploration of this line of thought.

## REFERENCES


Ahmed, Sarah. 2008. "Sociable Happiness." *Emotion, Space and Society* 1: 10–13.

Andre, Elizabeth. 2011. "Journeying through Despair, Battling for Hope: The Experience of One Environmental Educator." PhD diss., University of Minnesota.

Andreou, Chrisoula. 2007. "Morality and Psychology." *Philosophy Compass* 2: 46–55.

Bickman, L. 1972. "Environmental Attitudes and Actions." *Journal of Social Psychology* 87: 323–4.

Boler, Megan. 1999. *Feeling Power: Emotions and Education*. New York: Routledge.

Braithwaite, John. 2004. "Emancipation and Hope." *Annals of the American Academy of Political and Social Science* 592: 79–98.

Braithwaite, Valerie. 2004a. "Collective Hope." *Annals of the American Academy of Political and Social Science* 592: 6–15.

– 2004b. "The Hope Process and Social Inclusion." *Annals of the American Academy of Political and Social Science* 592: 128–51.

Brennan, Teresa. 2004. *The Transmission of Affect*. Ithaca: Cornell University Press.

Cafaro, Phil. 2005. "Thoreau, Leopold, and Carson: Toward an Environmental Virtue Ethics." In *Environmental Virtue Ethics*, edited by Ronald Sandler and Philip Cafaro, 31–44. Lanham: Rowman & Littlefield Publishers, Inc.

Campbell, Sue. 1994. "Being Dismissed: The Politics of Emotional Expression." *Hypatia* 9: 46–65.

– 1997. *Interpreting the Personal: Expression and the Formation of Feelings*. London: Cornell University Press.

Cheavens, Jennifer, Scott Michael, and C.R. Snyder. 2005. "The Correlates of Hope: Psychological and Physiological Benefits." In *Interdisciplinary Perspectives on Hope*, edited by Jaklin Eliott, 119–32. Hauppauge: Nova Science Publishers Inc.

Costanzo, M., D. Archer, E. Aronson, and T. Pettigrew. 1986. "Energy Conservation Behaviour: The Difficult Path from Information to Action." *American Psychologist* 41: 521–8.

Courville, Sasha, and Nicola Piper. 2004. "Harnessing Hope through NGO Activism." *Annals of the American Academy of Political and Social Science* 592: 39–61.

Cunningham, Kenneth. 2004. "An Autobiographical Approach to the Psychological Study of Hope." PhD diss., Institute of Clinical Social Work.

Cunsolo Willox, Ashlee, Sherilee L. Harper, Victoria L. Edge, Karen Landman, Karen Houle, James D. Ford, and the Rigolet Inuit Community Government. 2013. "The Land Enriches the Soil: On Climactic Environmental Change, Affect, and Emotional Health in Rigolet, Nunatsiavut, Canada." *Emotion, Space and Society* 6: 14–24.

Davidson Joyce, Liz Bondi, and Mick Smith. 2005. "Introduction: Geography's 'Emotional Turn.'" In *Emotional Geographies*, 1–16. Farnham: Ashgate Publishing Limited.

Drahos, Peter. 2004. "Trading in Public Hope." *Annals of the American Academy of Political and Social Science* 592: 18–38.

Durning, Alan. 2009. "How Much Is Enough?" In *Living Ethics: An Introduction*, edited by Michael Minch and Christine Weigel, 505–14. Belmont: Wadsworth, Cengage Learning.

Eliott, Jacklin. 2005. "What Have We Done with Hope? A Brief History." In *Interdisciplinary Perspectives on Hope*, edited by Jaklin Eliott, 3–45. Hauppauge: Nova Science Publishers Inc.

Fiala, Andrew. 2010. "Nero's Fiddle: On Hope, Despair and the Ecological Crisis." *Philosophy & the Environment* 15: 51–68.

Finger, M. 1994. "From Knowledge to Action? Exploring the Relationships between Environmental Experiences, Learning, and Behaviour." *Journal of Social Issues* 50, no. 3: 141–60.

Freire, Paulo. 2011. *Pedagogy of Hope: Reliving Pedagogy of the Oppressed*. New York: Continuum Publishing Company.

Frye, Marilyn. 1983. "A Note on Anger." In *The Politics of Reality: Essays in Feminist Theory*, 84–94. Trumansburg: The Crossing Press.

Geller, E.S. 1981. "Evaluating Energy Conservation Programs: Is Verbal Report Enough?" *Journal of Consumer Research* 8: 331–5.

Geller, E.S., J.B. Erickson, and B.A. Buttram. 1983. "Attempts to Promote Residential Water Conservation with Educational, Behavioural and Engineering Strategies." *Population and Environment Behavioural and Social Issues* 6: 96–112.

Goleman, Daniel. 2005. *Emotional Intelligence*. New York: Bantam Dell.

Goralnik, Lissy, and Michael Nelson. 2011. "Forming a Philosophy of Environmental Action: Aldo Leopold, John Muir, and the Importance of Community." *The Journal of Environmental Education* 42, no. 3: 181–92.

Harvey, David. 2007. *A Brief History of Neoliberalism*. New York: Oxford University Press.

Hsu, Shih-Jang. 2004. "The Effects of an Environmental Education Program on Responsible Environmental Behaviour and Associated Environmental Literacy Variables in Taiwanese College Students." *The Journal of Environmental Education* 35, no. 2: 37–48.

Hume, David. 2002. *A Treatise on Human Nature*. Edited by David Norton and Mary Norton. New York: Oxford University Press.

Hungerford, Harold, and Trudi Volk. 1990. "Changing Learner Behaviour through Environmental Education." *Journal of Environmental Education* 21, no. 3: 8–21.

Intergovernmental Panel on Climate Change. "Fifth Assessment Report: Summary for Policymakers." http://www.climatechange2013.org/images/uploads/WGI_AR5_SPM_brochure.pdf. Accessed 17 December 2013.

Jaggar, Alison. 1997. "Love and Knowledge: Emotion in Feminist Epistemology." In *Feminist Social Thought: A Reader*, edited by Diana Teitjens Meyers, 385–405. New York: Routledge.

Jasper, James. 1998. "The Emotions of Protest: Affective and Reactive Emotions in and around Social Movements." *Sociological Forum* 13: 397–424.

Jowit, Juliette. 2010. "Humans Driving Extinction Faster Than Species Can Evolve, Say Experts." http://www.guardian.co.uk/environment/2010/mar/07/extinction-species-evolve. Accessed 29 April 2013.

Kals, Elisabeth, Daniel Schumacher, and Leo Montada. 1999. "Emotional Affinity toward Nature as a Motivational Basis to Protect Nature." *Environment and Behavior* 31: 178–202.

Kelsey, Elin. 2003. "Constructing the Public: Implications of the Discourse of International Environmental Agreements on Conceptions of Education and Public Participation." *Environmental Education Research* 9: 403–27.

– 2012. "Do Not Despair." *New Scientist* 213: 24–5.

– 2013. "TEDxMonterey: Eco-buoyant!" http://www.youtube.com/watch?v=igVR2M6CMyM. Accessed 13 April 2013.

Kollmuss, Anja, and Julian Agyeman. 2002. "Mind the Gap: Why Do People Act Environmentally and What Are the Barriers to Pro-environmental Behaviour." *Environmental Education Research* 8, no. 3: 239–60.

Kretz, Lisa. 2012. "Climate Change: Bridging the Theory-Action Gap." *Ethics & the Environment* 17: 9–27.

– 2013a. "Hope in Environmental Philosophy." *Journal of Agricultural & Environmental Ethics* 26: 925–44.

– 2013b. "Conscious Evolution." *The Montréal Review*. http://www.themontrealreview.com/2009/Conscious-Evolution.php. Accessed 12 December 2013.

Kurtz, Lester. 2010. "The Mothers of the Disappeared: Challenging the junta in Argentina (1977–1983)." *International Center on Nonviolent Conflict*. http://www.nonviolent-conflict.org/index.php/movements-and-campaigns/movements-and-campaigns-summaries?sobi2Task=sobi2Details&sobi2Id=28. Accessed 5 May 2013.

Leopold, Aldo. 1949. *A Sand County Almanac: And Sketches Here and There*. New York: Oxford University Press.

– 1993. *Round River: From the Journals of Aldo Leopold*, edited by Luna Leopold. New York: Oxford University Press.

Lloyd, Genevieve. 1984. *The Man of Reason: "Male" and "Female" in Western Philosophy*. Minneapolis: University of Minnesota Press.

Lorde, Audre. 1984. *Sister Outsider: Essays and Speeches by Audre Lorde*. Berkeley: The Crossing Press.

Lueck, Michelle A.M. 2007. "Hope for a Cause as Cause for Hope: The Need for Hope in Environmental Sociology." *The American Sociologist* 38: 250–61.

McGeer, Victoria. 2004. "The Art of Good Hope." *Annals of the American Academy of Political and Social Sciences* 592: 100–27.

McIntosh, Peggy. 2000. "White Privilege and Male Privilege: A Personal Account of Coming to See Correspondences through Work in Women's Studies." In *Gender Basics: Feminist Perspectives on Women and Men*, 2nd edition, edited by Anne Minas, 30–8. Belmont: Thomson Wadsworth.

McKenzie-Mohr, Douglas. 2000. "Promoting Sustainable Behaviour: An Introduction to Community-Based Social Marketing." *Journal of Social Issues* 56, no. 3: 543–54.

Midgley, Mary. 1996. "Sustainability and Moral Pluralism." *Ethics & The Environment* 1: 41–54.

Monture-Okanee, Patricia. 1992. "The Violence We Women Do: A First Nations View." In *Challenging Times: The Women's Movement in Canada and the United States*, edited by Constance Blackhouse and David Flaherty, 193–200. Montreal: McGill-Queen's University Press.

Norgaard, Kari Marie. 2006. "'People Want to Protect Themselves a Little Bit': Emotions, Denial, and Social Movement Nonparticipation." *Sociological Inquiry* 76: 372–96.

Plumwood, Val. 2006. *Environmental Culture: The Ecological Crisis of Reason*. New York: Routledge.

Schwarzfeld, Matt. 2013. "Top 10 Ways Man Is Destroying the Environment." *Discovery*. http://dsc.discovery.com/tv-shows/curiosity/topics/10-ways-man-destroying-environment.htm. Accessed 2 May 2013.

Sherwin, Susan. 2008. "Whither Bioethics? How Feminism Can Help Re-orient Bioethics." *The International Journal of Feminist Approaches to Bioethics* 1: 7–27.

Shiva, Vandana. 2005. "The Impoverishment of the Environment: Women and Children Last." In *Environmental Philosophy: From Animal Rights to Radical Ecology*, edited by Michael Zimmerman, J. Baird Callicott, Karen Warren, Irene Klaver, and John Clark, 178–93. Upper Saddle River: Prentice Hall.

Sia, Archibald P., Harold R. Hungerford, and Audrey N. Tomera. 1985/86. "Selected Predictors of Responsible Environmental Behaviour: An Analysis." *Journal of Environmental Education* 17, no. 2: 31–40.

Smith, Mick. 2001. "Environmental Anamnesis: Walter Benjamin and the Ethics of Extinction." *Environmental Ethics* 23, no. 4: 359–76.

Snyder, Charles Richard. 1995. "Conceptualizing, Measuring, and Nurturing Hope." *Journal of Counseling & Development* 73: 355–60.
Spelman, Elizabeth. 1989. "Anger and Insubordination." In *Women, Knowledge, and Reality*, edited by Ann Garry and Marilyn Pearsall, 263–73. Winchester: Unwin Hyman.
Stotland, Ezra. 1969. *The Psychology of Hope: An Integration of Experimental, Clinical and Social Approaches*. San Francisco: Jossey Bass Inc.
Thoits, Peggy. 1989. "The Sociology of Emotions." *Annual Review of Sociology* 15: 317–42.
Thomashow, Mitchell. 1996. *Ecological Identity: Becoming a Reflective Environmentalist*. Cambridge, MA: MIT Press.
Thompson, Allen. 2010. "Radical Hope for Living Well in a Warmer World." *Journal of Agriculture and Environmental Philosophy* 23: 43–59.
Thrift, Nigel. 2004. "Intensities of Feeling: Towards a Spatial Politics of Affect." *Geografiska Annaler* 86 B: 57–78.
Turner, Jack. 1996. *The Abstract Wild*. Tucson: University of Arizona Press.
United Nations Report. 2006. "Livestock's Long Shadow – Environmental Issues and Options." http://www.fao.org/docrep/010/a0701e/a0701e00.htm. Accessed 2 May 2013.
Warren, Karen. 2005. "The Power and Promise of Ecofeminism, Revisited." In *Environmental Philosophy: From Animal Rights to Radical Ecology*, edited by Michael Zimmerman, J. Baird Callicott, Karen Warren, Irene Klaver, and John Clark, 252–79. Upper Saddle River: Prentice Hall.
Young, Iris Marion. 2009. "Social Movements and the Politics of Difference." In *Morality and Moral Controversies: Readings in Moral, Social and Political Philosophy*, edited by John Arthur and Steven Scalet, 582–8. Upper Saddle River: Pearson Education.
Zelman, Joanna. 2011. "Million Environmental Refugees By 2020, Experts Predict." *Huffington Post*. http://www.huffingtonpost.com/2011/02/22/environmental-refugees-50_n_826488.html. Accessed 13 April 2013.

# 11 Solastalgia and the New Mourning

GLENN ALBRECHT

## Introduction

As the global environment changes at an exponential rate, we need novel Earth-related concepts to give expression to our feelings about such change. I have created a psychoterratic (psyche-earth) typology of Earth-related emotions and feelings that captures the field as it has evolved in past and contemporary nature and place writing (Albrecht 2012a; Albrecht 2013). I have now located my concept of solastalgia (Albrecht 2005) within this psychoterratic typology. Solastalgia was created by me to fill a gap in our language of Earth-related emotions where people experience a deep form of existential distress when directly confronted by unwelcome change in their loved home environment. Solastalgia sits among closely related experiences of Earth-associated trauma, distress, grief, mourning, and melancholia that unfortunately are now often connected to escalating occurrences of acute and chronic environmental desolation. These negative Earth-related emotions and experiences afflict many hundreds of thousands of people worldwide as they are beset by unwelcome climate and landscape change in the Anthropocene (Crutzen and Stoermer 2000).

The futurist Alvin Toffler speculated in the early to mid-1970s that a combined economic and ecological "depression" that he saw as marking the end of growth-fixated industrio-technological societies would engulf the people of the Earth. In *Future Shock* (1970, 13), Toffler predicted epidemics of psychiatric disease and dislocation connected to the shock of rapid change. In *Eco-Spasm* (1975) he warned of "incomprehensible dread" at the collapse of the economic system and its associated energy and ecological foundations.

Since the global fiscal crisis of 2007, and with the combined impacts of global development pressures and global warming induced climate change, it is not unreasonable to suggest that Toffler was prescient in his prediction of connections between eco-economic depression and human depression. *The Diagnostic and Statistical Manual of Mental Disorders* (DSM) of the American Psychiatric Association (2013) defines the internationally accepted standard criteria and common language for classifying mental disorders. According to the system of categorization in the most recent edition, DSM-5, clinically diagnosable depression is on the rise worldwide with over 350 million people affected. In the late twentieth century clinically defined depression gradually replaced the older concept of melancholia as a specific mood disorder connected to a profound disturbance in a person's life. Nevertheless the debate is far from settled within psychiatry: melancholia was recently submitted to the DSM-5's mood-disorders committee to be added to conditions that could be biomedically diagnosed (Greenburg 2013).

In addition to medically defined depression and melancholia, there are other nonclinical forms of emotion, anxiety, and debilitating mental states that affect millions more people worldwide. In the context of the psychoterratic I take it as axiomatic that when – as Bateson (1973) argued – our physical landscape is being desolated, our mental landscape is equally disturbed. He argued in the context of the pollution of the Great Lakes of North America: "you decide that you want to get rid of the by-products of human life and that Lake Erie will be a good place to put them. You forget that the eco-mental system called Lake Erie is a part of *your* wider eco-mental system – and that if Lake Erie is driven insane, its insanity is incorporated in the larger system of *your* thought and experience" (Bateson 1973, 460).

As the scale of pollution ramps up to the global level, our eco-mental landscape includes the global climate and its associated weather events. Unfortunately, people in diverse parts of the world now have in common an anxiety due to the threat of increasing frequency and severity of extreme weather. Such "meteoranxiety" is exacerbated now that we have, 24/7, weather channels that deliver repeated, graphic warnings about every type of weather event possible.

In the field of non-medically diagnosable forms of mental dis-ease, I hope that the psychoterratic typology will help us to more clearly communicate the nuances of our feelings about these emergent psychological and emotional responses to changing environmental circumstances. I distinguish these existential and psychological states from somaterratic (body–earth) states (Albrecht et al. 2007; Albrecht 2012a). While the typology has both positive and negative components, it is the negatively perceived and felt environmental changes that are likely to prevail at this time in history when our home, the Earth, is in various forms of distress. From one of the first psychoterratic concepts, nostalgia, (Hofer [1688], 1934), to more recent contributions such as solastalgia (Albrecht 2005), these negative "Earth emotions" and psychological states are within the parameters of shock as predicted by Toffler.

As a consequence of living in the Anthropocene, the potential for negative psychoterratic experiences also reaches a planetary scope and scale within this epoch. Gone is the relative stability and predictability of the past 12,000 years as the established patterns and regularity of Holocene phenology begin to fall into chaos (Albrecht 2011). In the Anthropocene, the so-called "new normal," or what I prefer to conceptualize as "the new abnormal," will be characterized by uncertainty, unpredictability, genuine chaos, and relentless change. Earth distress – as manifest in global warming, changing climates, erratic weather, sea level rise, acidifying oceans, disease pandemics, species endangerment and extinction, bioaccumulation of toxins, and the overwhelming physical impact of exponentially expanding human development – will have its correlates in human psychoterratic and somaterratic distress.

While some have already felt the global dread (Albrecht 2012a) associated with the fear for the future we are creating for our descendants, others are directly experiencing the distress of actual change where the forces of development, rising sea levels, record storm surges, record heat, extreme weather, and record floods and droughts precipitate negative psychoterratic responses (Connor et al. 2004; Pereira 2008; McNamara and Westoby 2011; Berry et al. 2011; Cunsolo Willox et al. 2012; Cunsolo Willox et al. 2013; Smith-Cavros et al. 2012; Cordial et al. 2012). With the enhanced greenhouse effect and its attendant global warming now scientifically recognized threats to all humans and all other forms of life on this planet, there is an expanding domain of human emotion tied to the feelings of grief and loss at that which has already negatively changed or disappeared in the here and now. In addition, there is also anticipatory grief and mourning for that which is currently under stress and will most likely pass away in the foreseeable future. Phyllis Windle quotes from Bill McKibben's book *The End of Nature*, where he laments: "The end of nature probably also makes us reluctant to attach ourselves to its remnants, for the same reason that we usually don't choose friends from among the terminally ill. I love the mountain outside my back door ... But I know that some part of me resists getting to know it better – for fear, weak-kneed as it sounds, of getting hurt. If I knew as well as a forester what sick trees looked like, I fear I would see them everywhere. I find now that I like the woods best in winter, when it is harder to tell what might be dying. The winter woods might be perfectly healthy come spring, just as the sick friend, when she's sleeping peacefully, might wake up without the wheeze in her lungs" (McKibben, in Windle 1992, 364).[1]

## The New Mourning

The powerful grief felt when a loved one dies is a universal human experience, and even when the deceased person has lived a long and fulfilled life there is still intense sadness and grieving at their passing. Mortality and mourning go hand in hand. Traditionally, many cultures have expressed their grief at the loss of loved ones through signs and ceremonies that incorporate symbols of mourning. Religious ceremonies and many cultural traditions such as periods of social withdrawal, the raising of flags to half-mast, and the wearing of black apparel (including black arm bands) remain widespread in the world. Historically, in many cases, a component of the point and purpose of sacred and secular forms of mourning was to counter the apparent disorder, irrationality, and capriciousness of the way human life can be cut short in death (Morris 1987). In the contemporary world, there are still cultures where people reject the natural, scientific, causal explanations of sickness and death of modernity and, in their place, attribute power over life and death to forces such as sorcery and bad magic. "In many cultures, there is no cultural category of a 'natural death'; rather, each death is akin to murder in that a culturally sanctioned cause of death must be uncovered" (Abramovitch n.d.). In these circumstances the mourning ceremony could be used as a force exorcising the feelings of injustice and loss at the end of someone's life, especially if it was a life cut prematurely short by natural disaster or accident. In contemporary Papua New Guinea, for example, extreme instances of blame for death due to sorcery (sanguma) have become the subject of international attention where innocent people (mainly women) are caught up in violent waves of pathological grieving and mourning that inflict harm and even murder to others on the death of a close relative (Chandler 2013).

Despite the continuity and stability of the traditional elements of mourning, there are many emergent aspects of what could be called "the new mourning." I argue that the experience of grief at the loss of loved ones has conceivably changed for many people of any scientifically literate and educated culture. As opposed to past and some contemporary societies where often the direct cause of death was not known, humans now know the causes of most of the things that kill us. Even the power of invisible biological and geophysical forces like bacteria, viruses, and earthquakes is well understood by science. The task of an autopsy within the discipline of pathology is to give the precise cause of death to the relatives of the deceased on a death certificate. Earth scientists can give information about the precise strength, location, and depth of an earthquake to explain why one occurred in a particular place. The role of mourning then becomes less entangled in the mystery about who or what is directly responsible for the cause of death and is more closely tied to remembering the elements of the deceased life and how the abrupt cessation of interactions and intimate contact hurts those who remain. We no longer normally need to include interrelated feelings of grief and revenge about the cause of death as part of the mourning ceremony.[2]

The awareness of human culpability at a global scale is also a relatively new experience in the history of human mourning. Many humans now understand that we are often the primary agents of our own disasters. Environmental pollution, global warming, bad urban planning, and poor engineering are now major anthropogenic causes of human misery and death. In the Anthropocene there is no longer mystery attached to a great deal of disaster and misfortune since, to a very large extent, there is an element of self-imposed vulnerability to what are euphemistically called "natural disasters." We can now analyze human-enhanced natural disasters as part of what it means to be living in the Anthropocene. We choose to live on floodplains, we add carcinogens to our food supply, we burn fossil fuels to produce the particulates that cause lung disease, we add greenhouse gases to the atmosphere and cause global warming and its negative consequences for all humanity. It is this element of human agency with respect to the "hyperobject" of global warming (Morton 2013) that places an extra burden on our mourning (hyper-mourning) in the context of loss in both social and ecological contexts (Doherty and Clayton 2011).

In addition to knowledge of causality and human culpability altering the experience of mourning, grief and mourning about loss in the non-human realm are also emergent features of a globalized human culture. While people in traditional societies have undoubtedly grieved and mourned for the loss of non-human life, for some people, in the context of a scientific and technologically mediated world, such an experience might be novel. As the scientifically trained ecologist Phyllis Windle observed in the context of her perception of environmental decline in her poignant essay, "The Ecology of Grief": "Why, I wonder, did this bad news for the environment hit me so hard? Why do I want to commemorate the dying trees? I am an ecologist. Also, I am a trained hospital chaplain and chaplains are experts on death, dying, and grief. Finally, I realize: I am in mourning for these beautiful trees" (Windle 1992, 363).

Powerlessness in the face of pervasive change agents such as multinational corporations and authoritarian governments is another factor in the new mourning. Within the grip of such concentrated centres of power, individuals cannot direct their grief about negative environmental change towards anything or anyone in particular. Such powerlessness (personal and political) is also a defining feature of the concept of solastalgia. From the overview of aspects of grief and mourning above, I suggest that the new mourning contains the emergent elements of detailed knowledge of causality, anthropogenic culpability, and enhanced empathy for the non-human, as well as feelings of powerlessness.

## Mourning, Nostalgia, and Solastalgia

The etymological origins of the word "mourning" come from the Greek *memeros* related to "a state of being worried" and its meaning is associated with being troubled and to grieve. In addition, while in the contemporary context we see concepts such as melancholia (from the Greek for black bile and the Latin *lugere*, to mourn) as a form of sadness afflicting the psyche, in older times melancholia was also seen as something that could be manifest in physical symptoms of the body, hence the connection to the state of one's bile.

The melancholia associated with the original concept of nostalgia also had both physical and psychological dimensions. It was a trainee medical doctor, Johannes Hofer, who created the concept of nostalgia and published its detailed diagnosis in a dissertation written in Latin in Basel in 1688 (Hofer 1938). The neologism was a translation into Greek and New Latin of the German word *heimweh* or the pain for home – best translated into English as "homesickness." Similar concepts have been used in other languages for the feeling of loss when a person is separated from their much-loved home environment and wishes to return. Nostalgia (from the Greek *nostos* – return to home or native land – and the New Latin suffix *algia* – suffering, pain, or sickness from the Greek root *algos*) and/or the sickness caused by the intense desire to return home can be a source of profound and interrelated psychological and physiological distress.

According to Hofer the symptoms of nostalgia included a whole range of psychological and bodily afflictions ranging from intense sadness to palpitations of the heart. Hofer suggested that nostalgia was most likely to be experienced by people who were forcibly removed from their home environment such as soldiers transported to fight on foreign soil. In addition to war, other forms of prolonged absence from loved home environments were a likely cause of nostalgia. Nostalgia, as Hofer defined it, was still present in medical diagnosis until the mid-twentieth century but the term gradually became associated with a sentimental sense of regret and longing for places or periods in the past where a person felt more at ease. This form of sentimentality is no longer associated with profound psychological melancholia or physical illness and is no longer discussed in medico-psychological literature as a diagnosable illness. However, as I have argued elsewhere (Albrecht 2005, 2011), there is still a case for the relevance of Hoferian nostalgia in contexts where people closely tied to traditionally occupied homeland are forcibly removed.[3] Such forced removals are a well-documented part

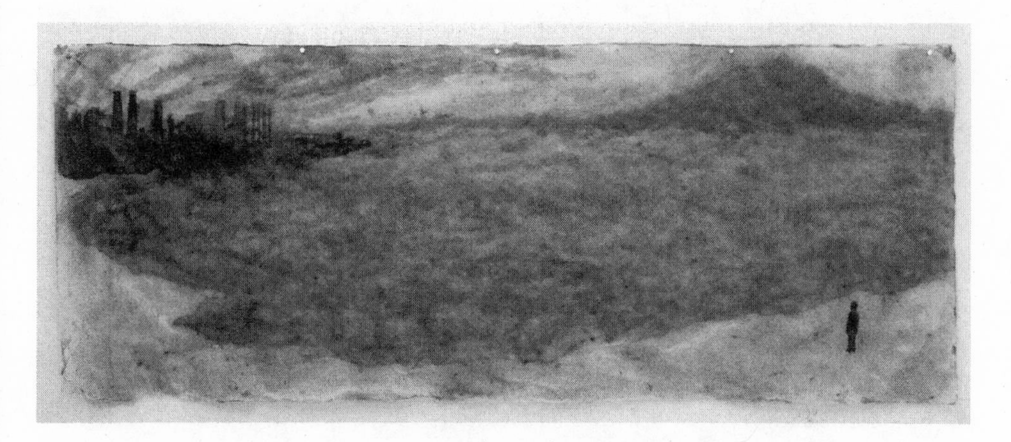

Figure 11.1 | Michelle Wilson, *Solastalgia*. michellewilsonprojects.com.

of the colonial experience of Indigenous people in places like Australia and North America, and in parts of the Amazon, the Central Kalahari, Indonesia, Tibet, and China where people are still being moved en masse against their will. Once removed, their genuine nostalgia for their former homeland would be palpable and must rank as a major source of psychoterratic and somaterratic distress (psychosomaterratic distress).

The concept of solastalgia (Albrecht 2005, 2011, 2012a, b) captures the chronic distress and melancholia of the lived experience of negative environmental change. It is as if one is living with a loved person who is seriously ill and who is suffering. There is a grieving for the loss of the signs and symbols of a healthy ecosystem and place. There is distress about the degraded state of the environment. The concept was created as a result of my personal engagement with distressed people and their desolated landscapes in the Hunter Valley of New South Wales in Eastern Australia. In this case the negative change to their environment was caused by open-pit or open-cut coal mining and the attendant infrastructure needed to mine and transport the coal. Solastalgia is defined as a particular form of psychoterratic distress connected to negatively perceived and felt changes to a home environment, changes that one is powerless to prevent. Solastalgia is succinctly described as "the homesickness you have when you are still at home."

Unlike medically defined nostalgia the context within which solastalgia was defined was the "existential" or lived experience of negative environmental change (Connor et al. 2004; Albrecht 2005). Solastalgia was created to fill a void in the English language for a non-medically defined psychoterratic emotion. In other words, solastalgia is not an illness that can be diagnosed and then treated by the biomedical sciences of medicine or psychiatry. It is a form of distress or grief for the lived experience of the desolation of one's home environment (on any scale).

The concept of solastalgia has its etymological origins in the concepts of nostalgia, solace, and desolation. Solace is derived from the Latin verb *solari* (noun *solacium*), with meanings connected to the alleviation or relief of distress or to the provision of comfort or consolation in the face of distressing events. Special environments, especially those that provide topophilia (Tuan 1974; McManus et al. 2013), might provide solace in ways that other places cannot. If a person lacks solace then they are in need of consolation. If a person tries to seek solace from or solitude in a much-loved place that is being desolated, then they will suffer distress.

Desolation has its origins in the Latin *solus* (noun *desolare*) with meanings connected to devastation, deprivation of comfort, abandonment, and loneliness. It, like solace, has meanings that can relate to both psychological and physical contexts, to a personal feeling of abandonment (isolation) and to a landscape that has been devastated. A desolated landscape can produce feelings of desolation in people. In addition, I also constructed the concept of solastalgia so that it has a reference and structural similarity to the concept of nostalgia, which has a place/home reference embedded. Hence, although solastalgia has its origins in the New Latin word "nostalgia" (and its Greek roots *nostos* and *algos*) it is based on two Latin roots, "solace" and "desolation," with the New Latin suffix algia, or pain, to complete its meaning.

Solastalgia can be contrasted to both the positive psychoterratic feelings associated with love of place, or topophilia, and also to the negative feelings of spatial and temporal dislocation and dispossession experienced as traditionally defined nostalgia. Solastalgia defines the existential, lived experience of the loss of value in the present as manifest in a feeling of disorientation, of being undermined by forces that destroy the potential for solace to be derived from the home environment.

The most common circumstances where solastalgia is likely to be evident are with chronic and pervasive change to the existing biophysical order. In cases such as coal mining or gas fracking, the challenge to one's sense of place is unrelenting and these activities can be present over decades or longer in the home environment and irreversibly alter the landscape. Changes to the built environment can also be a cause of solastalgia. With the ongoing demolition of older and long-lived buildings and their replacement with contemporary short-lived and rapidly constructed buildings, urban and city landscapes can transform and, along with that change, elicit solastalgia, especially in citizens who have lived a long time in the former, more stable, urban environment. In addition, all humans now experience changing home environments due to global warming and as these changes become more extreme and more pervasive, more people will be negatively affected in both physical and mental health contexts (Swim et al. 2011; Albrecht 2011; Berry et al. 2010; Cunsolo Willox 2012; Doherty and Clayton 2011). Moreover, "we can experience solastalgia when we erode our relationships with other beings" (Bekoff 2007, 163) so our "home" includes those non-human beings who share an environment with us.[4]

The associations between grief, the new mourning, melancholia, solastalgia, and nostalgia form a subset of elements of a much more complex and diverse set of conditions that can be applied to the way people are affected by changes to their home environment. There are many more nuances of negative psychoterratic emotional states and we often struggle to give adequate expression to the emotions of grief and loss. This is particularly so when our emotions are about special, loved places or beloved non-human "friends" that are becoming rare or have gone locally extinct. Such ill-defined feelings include the instant distress when confronted by rapid and destructive environmental change (such as the felling of a historic and significant tree) and the anticipatory dread associated with no longer wanting to visit and see places you once knew and loved, that you now know have been utterly transformed for the worse. Given the likely increase in these negative feelings in an era of expansive, exponential change, it is timely to give full expression to our negative psychoterratic emotions and to do so in such a way that enables sharing and education. It is the emotions that move us, and, without a language with which to give carriage to our earthly feelings and emotions, we are distressed as well as impoverished and disempowered. With ownership of such a language and conceptual framework, I argue we are not only more articulate; we are more empowered to bring about change towards the conditions of life that support and nurture positive mental and physical health.

## The Psychoterratic Typology: The Struggle between Terranascia and Terraphthora

As indicated above, in order to understand the many nuances of our emotional responses to our environment, I have created a typology of psyche-Earth emotions and feelings that have been developed by myself and scholars in this transdisciplinary field since 2003. I see the full range of these psychoterratic emotions as sitting between the extremes of positive and negative responses. In the spirit of systematically "re-placing" our language I have decided to give to these poles of experience names that suitably define the terrain. The full range of the psychoterratic can be circumscribed within the poles of Terraphthora (Earth destroyer) (*tera for ra*) (*Terra*, from the Latin "earth," and the Greek *phthorá* or "destruction") and Terranascia (Earth creator) (*tera nas cia*) (*Terra*, from the Latin "earth" and the Latin *nātūra*, "to be born"). The dialectic between Terraphthora and Terranascia is now being expressed in all forms of human creativity and destruction and it is my hope that the typology as a whole can assist in the ecocritical evaluation of ideas and actions that relate to our home environment at all scales. As I shall argue in more detail below, the explicit giving of names[5] to that which had previously been intangible or subliminal is empowering, and enables those who participate in the named drama to be collaboratively creative and to engage in a community of scholarship, politics, and criticism.

In addition to the creation of new terms to encompass and frame our new situation, it may be necessary to stop using terms that have become abused or meaningless. One such central term in the person-place lexicon is the concept of the environment. If the environment is defined as that which surrounds us, that which is outside, then we have already made a category mistake. This is in part because we ourselves provide an environment for the organisms that live within us. We are the basis for life inside us and there is interchange between the inside and the outside through endless permeability. We are reminded of this permeability when we get diseases such as influenza, malaria, or food poisoning. Nevertheless if you look closer you will find a whole ecosystem of organisms living in and on your own cell tissues. Both "good" and "bad" bacteria are rushing in and out of us in their trillions via the media of solids, gases, and fluids. Tiny eyelash mites inhabit our eyelash follicles and eat dead cells. At the microscopic scale, there is no barrier between us and the outside ... it is all one seething, interrelated party where ... as they say in the classics, "everything is connected to everything else." Of course, the same applies at the macroscopic scale where humans are embedded within what we might otherwise term the environment.

To avoid such a category mistake, I have decided that we actually *all* live in the "symbionment." This term has its origins in the word "symbiosis," which in turn is from the Greek *sumbiōsis* (companionship), *sumbioun* (to live together), and *sumbios* (living together). Within the symbionment we live together with relatively harmonious companionship or a state of ecosystem health between ourselves and other beings. I have argued elsewhere that ecosystem health and ethical goodness can be seen as mutually supportive and such living together can be the foundation of ideas of good health and the ethically good (Albrecht 2001).

While we are at it, since symbiosis is such a central feature in biology (life), then we need to get rid of the concept of the Anthropocene before it covers many more decades of the history of Earth. The distress of the Earth is being driven by human induced failures in many domains including ecosystem obliteration (development), ecosystem distress syndromes, climate chaos, extinction cascades, forced species relocation, emergent diseases, and over-reach, then failure, of vital biogeochemical cycles. I want this period in history to become redundant as soon as possible, since, the longer it prevails, the more likely we will suffer catastrophic failure as a species here on Earth. While this would be a tragedy of huge proportion for humans, we will also take with us thousands, perhaps millions, of other species as well. Popular literature and film already portray such an apocalyptic turn in human-nature relationships. In the dying moments of the late Anthropocene, where a small number of humans continue to eke out a tortuous existence, they will no doubt experience profound grief and the "new mourning" for the lost vitality of the past.

In order to counter this negative trend in our cultural responses to the current crises, we clearly need, within popular culture, visions and memes of a terranascent future. To get the detail into such visions, we will need yet more new conceptual development, since the foundation on which we are building right now is seriously flawed and conducive of nothing but great waves of ennui, grief, dread, solastalgia, mourning, and melancholia. The new foundation, built around the new meme of the Symbiocene (see below), will need to be an act of positive creation. It is my hope that by naming and outlining such a desirable future, young people will be inspired to give it content.

While we have already tried to build a new and viable society around concepts such as sustainability, sustainable development, and resilience, all these terms have been corrupted by forces determined to incorporate and embed them into the Anthropocene where they become "business as usual." "Sustainability" is inadequate as a concept because it does not specify what is to be sustained and over what time frame it is to be sustained. Equally, sustainable development fails to define what it is about development that is to be sustained ... except perhaps development itself (Albrecht 1994). Yet global-scale development that is not broadly in harmony with greater life and planetary forces puts us on the path to dislocation and ultimately extinction.

The concept of resilience (Walker and Salt 2006) has also been appropriated by forces determined to pull it into the gravitational influence of industrial society on a globalized scale. Instead of helping us rebound into configurations of successful models of living after disturbance, we are now seeing resilience being used to justify the ongoing existence of processes and activities that are driving humans to extinction. Coal, oil, and gas fracking industries now use their public relations departments to spin the message that their industries are not only sustainable, but resilient as well. The ongoing resilience of technically non-sustainable and undesirable features of social systems is more correctly termed "negative resilience" (Gallopín 2006) or "perverse resilience" (Holling 2001; Ráez-Luna 2008). These forms of resilience occur where pathological social relationships that are oppressive and exploitative of humans and ecosystems are rendered resistant to change by economic and political subsidies (donations), political support, and vested interests.

## Entering the Symbiocene: Sumbiosity and Sumbiosic Development

The ecological process known as symbiosis is critical to the long-term viability of life on this planet. As a scientific term, it has been used to give substance to the nature of the interactions between different organisms living in close physical association. The relatively recent discovery of immense associations of macrofungi and flowering plants in symbiotic relationship to each other in ecosystems all over the world has already overturned the dominance of the view of life as solely founded on competitive struggle between species (Albrecht 2001). We now know that, for example, health in a forest ecosystem is regulated by what are called "mother trees," which control fungal networks that interconnect trees of varying ages and regulate nutrient flows to those trees that need them (Simard et al. 2015).

Ecological complexity via the emergence of symbiosis gives interconnectedness in space (spatial relationships) and evolution by natural selection gives us interconnectedness in time (temporal connectedness). Between the two, we get a unified account of how life is possible and how it has both directionality and continuity (Albrecht 1998). As argued by Kropotkin and, more recently, by Murray Bookchin in his 1964 essay "Ecology and Revolutionary Thought," ecology is a revolutionary domain that requires us to see the interconnections between the elements in complex adaptive systems (biogeochemical systems) as *foundational* for all life.[6] Environmental philosophers such as Aldo Leopold simply expressed the idea of interconnectedness as a form of mutualism where we all live within the "biotic community." While predator-prey relationships exist within the biotic community, it is only possible for continuity of the community if relative "balance" within all the relationships prevails. Such a foundation for life was the main theme of Leopold's famous essay in *A Sand County Almanac*, "Thinking like a Mountain." Too many predators and the system collapses, too many herbivores and even mountains can crumble away to nothing as vegetation is removed and erosive powers prevail (Leopold 1966).

## The Symbiocene

The concept of symbiosis is also the foundation for the concept of the symbionment and the period of history I call the Symbiocene, the era that we must enter as we rapidly move out of the Anthropocene. The new era will be characterized by human intelligence that replicates the symbiotic and life reproducing processes to be found in life and natural systems. The elements include full recyclability of all inputs and outputs, safe and socially just renewable energy (preferably produced from that safe and distant nuclear reactor, the sun), full and harmonious integration with physical and living systems at all scales, and the elimination of waste in all aspects of human enterprise. It is within such an era that solastalgia, grief, and the new mourning will also be minimized as, although we cannot negate the aging process and avoid death, we can prevent those anthropogenic actions that bring about our own premature morbidity and mortality. We will no longer have to mourn our own culpability and stupidity as well as natural mortality.

These considerations also apply to our relationships with non-human life. It is now possible to more fully appreciate that non-human animals also grieve and mourn for those close to them that die (Bekoff 2007). Humans can empathize with non-humans that are caught up in negative and life threatening change that has been imposed upon them, so much so that we might even wish to intervene in their lives and move them to "safer" places (Albrecht et al. 2012). If we can create the Symbiocene, all forms of grieving and mourning for non-humans will be reduced in frequency and intensity as we rebuild our positive relationships with them. Within the social lives of non-human beings, forms of grief and mourning will also continue for "normal" loss within life cycles, but will not be associated with gross negative change as imposed by humans.

So that we might transition into the Symbiocene, we will need to eschew concepts that have been illegitimately appropriated by those in control of the Anthropocene. In order to (re)discover the symbionment, we will need to replace sustainability and sustainable development as concepts. I suggest "sumbiosity" and "sumbiosic development" as new terms that will help us get out of the Anthropocene. Sumbiosity[7] I define as an emergent state actively created by humans in all domains that conserves and/or achieves an ongoing balance and permanent interdependence between humans, all other life forms (e.g. mycorrhizal associations, predator-prey relationships) and the life support systems of the Earth (e.g. biogeochemical cycles). This concept has the same roots as symbiosis except that the emphasis is on the root Greek word *sumbios*, or "living together." Sumbiosity is a state of living together within a "biotic community" (Leopold 1989) that achieves an ongoing ecological balance with other life forms of the Earth (a state formerly described as sustainability).

All of the essential characteristics of the Symbiocene, as outlined above, will be described as "sumbiosic." Sumbiosic human actions and enterprise will be exemplified by those cumulative types of active and purposive relationships and attributes created by humans that enhance mutual interdependence and mutual benefit for all living beings so as to conserve and maximize a state of sumbiosity. Hence, sumbiosic development will consist of creative development that uses the very best of biomimicry and "symbiomimicry"[8] plus other eco-industrial, eco-technological, and eco-agricultural innovation to ensure that human societies live within the biotic community and not in opposition to it.[9] Sumbiosic development will be a form of development that has sumbiosity as its goal or end state and will satisfy the condition of living within the Symbiocene.

## Expanding the Psychoterratic Typology

In order to get out of the Anthropocene and into the Symbiocene, we need to fully appreciate the positive and negative aspects of our emotional relationships to land, landscape, and ecosystems. The dramatic struggle between Terranascia and Terraphthora is being played out right now on the stage of the whole Earth. Where terraphthoric forces are winning, there are negative psychoterratic emotions including grief and mourning for life that is lost, endangered, displaced, and disturbed. Where the terranascent forces are winning, the celebration of life and living processes is characterized by a set of positive emotions.

We can more fully appreciate the drama between the forces of destruction and creation if we examine some emergent elements of the psychoterratic typology. The typology is still evolving as new concepts are being added both by me and by others interested in mind-Earth relationships. While detailed definitions and a table indicating all known historical concepts have been provided in my previous publications (see, for example, Albrecht 2012a, 2013), I will define some new terms created by me and others as they will not be in any dictionary or previous print publications (in English). The newly defined terms that must be placed in the negative or terraphthoric side of the typology include my concepts of "topoaversion" and "tierratrauma," as well as Heneghan's idea of "toponesia" (Heneghan 2013).

Topoaversion is a conceptual response to the feeling that you do not wish to return to a place that you once loved and enjoyed when you know that it has been irrevocably changed for the worse. It is not topophobia, where you have fear of a place while entering it; topoaversion is a strong enough feeling to keep you from ever returning to visit the place that was once beloved. The concept has its origins in topos (place), and aversion (to turn away).

Tierratrauma describes that moment when you experience sudden and traumatic environmental change ... your favourite tree is being cut down, fire is destroying your local area, a bulldozer is demolishing your loved streetscape, or you are witness to an oil spill that smothers all life on "your" beach. This is not medically defined post-traumatic stress disorder; this is not solastalgia derived from chronic change; this is acute Earth-based existential trauma. We needed a concept for this commonly felt, gut-wrenching mental anguish. Tierratrauma can be contrasted to eutierria (Albrecht 2012a), a concept I have already offered as an earthly catch-all for what has previously been called an oceanic experience or euphoric feeling of oneness or indivisibility between self and nature. It can also be opposed to endemophilia or the love of that which is distinctive (unique) about a particular place. If we can experience the pleasures of topophilia and eutierria, then we can also have tierratrauma when we witness and experience the sudden destruction of that which we love in our landscape.

Toponesia describes the process of forgetfulness of precious places that afflicts us as we leave the world of our childhood (Heneghan 2013). Heneghan writes powerfully about place, as explicated in the story of Winnie the Pooh: "But there is another sadness recorded in Christopher Milne's story, a sadness that most of us experience, I expect: the loss of connection with place, especially a natural one, that happens as we grow older. I propose, in the spirit of Albrecht, to call this 'toponesia' (from the Greek *topos*, place, and *amnesia*, loss of memory). Even if the world stood still, we would still spin away from it, dragged into the orbit of our private economies and that series of mischiefs that we call our adult life. These psychological factors associated with *Winnie-the-Pooh* – its nostalgia, solastalgia and toponesia – combine to make the stories a surprisingly powerful meditation on place, as much as a source of simple pleasure" (Heneghan 2013).

## Concluding Remarks

To mourn is both a biologically and culturally mediated human (and other animal) experience and we have now entered a new phase of global-scale mourning. As argued above, the emergent elements of death and mourning include reliable knowledge of causality, the role of human culpability, empathy for loss of life in the non-human realm, and feelings of powerlessness. All four elements alter the experience of death, grief, and mourning in a globalized human culture. Perhaps we should expand the psychoterratic typology beyond an established term such as "ecoanxiety" to include a concept like "memerosity" or what I would define as the pre-solastalgic state of being worried about the possible passing of the familiar and its replacement by that which does not sit comfortably within one's sense of place. I begin to mourn for that which I know will become endangered or extinct even before these events unfold. I often have a tight knot of memerosity inside me when I consider the scale of change going on around me, and what might happen next.

Not only do we have global dread (Albrecht 2012) about the future but since a great deal of negative change has already occurred on our planet, mourning for that which has already gone or is under intense stress is a perfectly reasonable response. We are entering an age of solastalgia (Albrecht 2012b). If we go even deeper into the Anthropocene we can see the potential for psychoterratically induced melancholia, grief, and the emergence of hyper-mourning.

As the ancient Greeks knew all too well, the psychoterratic drama is enhanced when disaster is self-inflicted. Yet it is our ability to foresee such potential for negativity that has prompted me to create a positive vision of the Symbiocene with new, positive psychoterratic concepts that are the direct opposite of those in the Anthropocene. People whose emotions and thinking allow them to already feel part of the Symbiocene will have the ability to grieve and, where appropriate, to mourn the loss of individual life (all beings) and the vitality of ecosystems as

Figure 11.2 | Michelle Wilson, *Mourning the World*. michellewilsonprojects.com.

negative change impinges on the planet. These people will also have the capacity and creative energy to transcend the Anthropocene and help others to enter the Symbiocene.

One of the first journalists to write about the concept of solastalgia for an international audience was Clive Thompson in *Wired* magazine (Thompson 2007). The headline for the story was "Global Mourning: How the Next Victim of Climate Change Will Be Our Minds." In a neat journalistic twist on the idea of global warming, Thompson was able to help readers see the connection between climate change and global mental health in an instant. It is my hope that with new contributions to the language of our Earth and Earth emotions, this chapter will sit with the others in this book as, first, a way of understanding our own reactions to the change that is occurring in and around us and, second, a way out of negative Earth emotions and into a new world of positive Earth emotions. Global solastalgia and the new mourning must give way to tierraphilia and the new exuberance for life.

NOTES

1 I dedicate this essay to my mother, Thelma Edwards, who passed away due to respiratory distress while I was writing about death and mourning

for this anthology. It was my mother who provided me with such a strong endemic sense of place, and thus my profound sadness when an endemic sense of place was lost.

2 Where the cause of death is murder or some other form of culturally or legally proscribed behaviour, feelings of grief and revenge are likely to be tied to the process of mourning.

3 Moreover, there is a case for recognizing the synergistic interaction of the psychoterratic and the somaterratic in the account of dis-ease. The body and the psyche do not exist separately in some unfathomable Cartesian divide (Albrecht 2011).

4 I have also speculated that climate stress on non-human beings (especially sentient animals) might cause them to experience emotions similar to solastalgia as their home habitat changes around them and they respond to the imperative to move (Albrecht 2011, 44–5).

5 While such "name giving" might appear to be an example of the social construction of reality I argue that this naming is a legitimate response to novel changes in the biophysical world, outside of cultural and evolutionary experience and in areas where language has not previously had any traction. In the same way, I do not see the novel emotional responses in people to such pervasive change in terms of reductionism, where emotional reactions to change are pathologized and seen as "mental illness." On the contrary, I see these novel Earth emotions as rational responses to irrational situations imposed on people. Examples of such novel Earth emotions occur when, for example, open pit coal mining or gas fracking forcibly enter the territory and homes of rural and remote people.

6 Such a view does not entail a steady state or climax community as the inevitable outcome of ecological processes. As I have argued elsewhere (Albrecht 2001), disturbance and change are vital elements of ecological processes generating diversity within complex adaptive systems; however, too much and too frequent disturbing change leads to the impossibility of robust symbiotic and resilient adaptive processes and hence system failure. It is precisely too much change to life support systems (boundary conditions) that is leading us into planetary crisis (Rockstrom et al. 2009).

7 Pronounced "soom-bi-os-ity."

8 A neologism to describe a field of creativity and design that has yet to be invented.

9 I wish to leave the detail of the creation and content of the Symbiocene for a future publication; however, I feel that this idea could be taken up by those (of any age) who wish to use their terranascent creativity and optimism to make manifest that which I can only see as an idea. After all, I am a philosopher, not an artist, engineer, designer, or architect.

REFERENCES

Abramovitch, H. n.d. “Anthropology of Death.” http://henry-a.com/apage/115917.php. Accessed 18 December 2013.

Albrecht, Glenn A. 1994. “Ethics, Anarchy and Sustainable Development.” *Anarchist Studies* 2, no. 2: 95–118.

– 1998. “Ethics and Directionality in Nature.” In *Social Ecology after Bookchin*, edited by Andrew Light, 83–92. New York: Guilford Press.

– 2001. “Applied Ethics in Human and Ecosystem Health: The Potential of Ethics and an Ethic of Potentiality.” *Ecosystem Health* 7, no. 4: 243–52.

– 2005. “Solastalgia: A New Concept in Human Health and Identity.” PAN (*Philosophy, Activism, Nature*) 3: 41–55.

– 2011. “Chronic Environmental Change and Mental Health.” In *Climate Change and Human Well-Being: Global Challenges and Opportunities*, edited by Inka Weissbecker, 43–56. New York: Springer.

– 2012a. “Psychoterratic Conditions in a Scientific and Technological World.” In *Ecopsychology: Science, Totems, and the Technological Species*, edited by Peter H. Kahn Jr and Patricia H. Hasbach, 241–64. Cambridge, MA: MIT Press.

– 2012b. “The Age of Solastalgia.” *The Conversation*. http://theconversation.com/the-age-of-solastalgia-8337. Accessed 7 August 2012.

– 2013. “Solastalgie: Heimweh in der Heimat.” In *Natur im Blick der Kulturen: Naturbeziehung und Umweltbildung in fremden Kulturen als Herausforderung für unsere Bildung. Eberswalder Beiträge zu Bildung und Nachhaltigkeit, Bd.1*, edited by Norbert Jung, H. Molitor, and A. Schilling, 47–60. Opladen: Budrich Uni-Press.

Albrecht, Glenn A., C. Brooke, D. Bennett, and S.T. Garnett. 2012. “The Ethics of Assisted Colonization in the Age of Anthropogenic Climate Change.” *The Journal of Agricultural and Environmental Ethics* 26, no. 4: 827–45.

Albrecht, Glenn A., Gina Sartore, Linda Connor, Nick Higginbotham, Sonya Freeman, Brian Kelly, H. Stain, A. Tonna, and G. Pollard. 2007. “Solastalgia: The Distress Caused by Environmental Change.” *Australasian Psychiatry* 15: 95–8.

American Psychiatric Association. 2013. *Diagnostic and Statistical Manual of Mental Disorders.* 5th edition. http://www.psych.org/practice/dsm. Accessed 20 December 2013.

Bateson, Gregory. 1973. *Steps to an Ecology of Mind.* London: Granada Publishing Limited in Paladin Books.

Bekoff, Marc. 2007. *The Emotional Lives of Animals*. California: New World Library.

Berry, Helen, Kathryn Bowen, and Tord Kjellstrom. 2010. “Climate Change and Mental Health: A Causal Pathways Framework.” *International Journal of Public Health* 55: 123–32.

Berry, Helen, A. Hogan, J. Owen, D. Rickwood, and L. Fragar. 2011. "Climate Change and Farmers' Mental Health: Risks and Responses." *Asia-Pacific Journal of Public Health* 23: 1295–1325.

Bookchin, M. 1986. "Ecology and Revolutionary Thought." In *Post-Scarcity Anarchism, Second Edition*. Montreal: Black Rose Books.

Chandler, Jo. 2013. "It's 2013, and They're Burning Witches." *Globe and Mail*, 15 February. http://www.theglobalmail.org/feature/its-2013-and-theyre-burning-witches/558/. Accessed 21 December 2013.

Connor, Linda, Glenn Albrecht, Nick Higginbotham, Sonia Freeman, and Wayne Smith. 2004. "Environmental Change and Human Health in Upper Hunter Communities of New South Wales, Australia." *EcoHealth* 1, no. 2: 47–58.

Cordial, Paige, Ruth Riding-Malon, and Hilary Lips. 2012. "The Effects of Mountaintop Removal Coal Mining on Mental Health, Well-Being, and Community Health in Central Appalachia." *Ecopsychology* 4, no. 3: 201–8.

Crutzen, P.J., and E.F. Stoermer. 2000. "The 'Anthropocene.'" *International Geosphere-Biosphere Program Newsletter* 41: 17–18.

Cunsolo Willox, Ashlee. 2012. "Climate Change as the Work of Mourning." *Ethics and the Environment* 17, no. 2: 137–64.

Cunsolo Willox, Ashlee, Sherilee Harper, Victoria Edge, Karen Landman, Karen Houle, James D. Ford, and the Rigolet Inuit Community Government. 2013. "'The Land Enriches the Soul': On Environmental Change, Affect, and Emotional Health and Well-Being in Nunatsiavut, Canada." *Emotion, Space, and Society* 6: 14–24. doi:10.1016/j.emospa.2011.08.005.

Cunsolo Willox, Ashlee, Sherilee Harper, James D. Ford, Victoria Edge, Karen Landman, Karen Houle, Sarah Blake, and Charlotte Wolfrey. 2013. "Climate Change and Mental Health: An Exploratory Case Study from Rigolet, Nunatsiavut, Labrador." *Climatic Change*. doi.:10.1007/s10584-013-0875-4.

Cunsolo Willox, Ashlee, Sherilee Harper, James D. Ford, Karen Landman, Karen Houle, Victoria Edge, and the Rigolet Inuit Community Government. 2012. "'From This Place and of This Place': Climate Change, Health, and Place in Rigolet, Nunatsiavut, Canada." *Social Sciences and Medicine* 75, no. 3: 538–47.

Doherty, Thomas, and Susan Clayton. 2011. "The Psychological Impacts of Global Climate Change." *American Psychologist* 66, no. 4: 265–76.

Gallopin, Gilberto C. 2006. "Linkages between Vulnerability, Resilience and Adaptive Capacity." *Global Environmental Change* 16: 293–303.

Greenburg, Gary. 2013. "Does Psychiatry Need Science?" *New Yorker*, 23 April. http://www.newyorker.com/online/blogs/elements/2013/04/psychiatry-dsm-melancholia-science-controversy.html. Accessed 24 November 2013.

Heneghan, Liam. 2013. "The Ecology of Pooh." http://www.aeonmagazine.com/nature-and-cosmos/liam-heneghan-ecology-childhood/. Accessed 24 November 2013.

Hofer, Johannes. 1934. "Medical Dissertation on Nostalgia." *Bulletin of the Institute of the History of Medicine* 2, no. 6: 376–91.

Holling, C.S. 2001. "Understanding the Complexity of Economic, Ecological, and Social Systems." *Ecosystems* 4: 390–405.

Kropotkin, Peter. 1987. *Mutual Aid: A Factor in Evolution*. London: Freedom Press.

Leopold, Aldo. 1989. *A Sand County Almanac*. New York: Oxford University Press.

McManus, Phil, Glenn Albrecht, and Raewyn Graham. 2014. "Psychoterratic Geographies of the Upper Hunter Region, Australia." *Geoforum* 51: 58–65. http://dx.doi.org/10.1016/j.geoforum.2013.09.020.

McNamara, K.E., and R. Westoby. 2011. "Local Knowledge and Climate Change Adaptation on Erub Island, Torres Strait." *Local Environment: The International Journal of Justice and Sustainability* 16, no. 9: 887–901.

Morris, Ian. 1987. *Burial and Ancient Society*. New York: Cambridge University Press.

Morton, Timothy. 2013. *Hyperobjects: Philosophy and Ecology after the End of the World*. Minneapolis: University of Minnesota Press.

Pereira, R.B. 2008. "Population Health Needs beyond Ratifying the Kyoto Protocol: A Look at Occupational Deprivation." *Rural and Remote Health* 8: 927.

Ráez-Luna, E. 2008. "Third World Inequity, Critical Political Economy, and the Ecosystem Approach." In *The Ecosystem Approach – Complexity, Uncertainty, and Managing for Sustainability*, edited by David Waltner-Toews, J.J. Kay, and N. Lister. New York: Columbia University Press.

Rockström, Johan, Will Steffen, Kevin Noone, Åsa Persson, F. Stuart Chapin, Eric F. Lambin, Timothy M. Lenton, et al. 2009. "A Safe Operating Space for Humanity." *Nature* 461: 472–5.

Simard, Suzanne, Amanda Asay, Kevin J. Beiler, Marcus A. Bingham, Julie R. Deslippe, Xinhua He, Leanne J. Philip, Yuanyuan Song, and François P. Teste. 2015. "Resource Transfer between Plants through Ectomycorrhizal Networks." In *Mycorrhizal Networks*, edited by T.R. Horton. Springer, forthcoming.

Smith-Cavros, E., Sylvia Duluc-Silva, Maria del Carmen Rodriguez, Ponciano Ortiz, and Edward O. Keith. 2012. "'You Can't Eat Money When You Are Hungry': Campesinos, Manatee Hunting, and Environmental Regret in Veracruz, Mexico." *Culture, Agriculture, Food and Environment* 34, no. 1: 68–80.

Swim, Janet, Paul Stern, Thomas Doherty, Susan Clayton, Joseph Reser, Elke Weber, Robert Gifford, and George Howard. 2011. "Psychology's Contributions to Understanding and Addressing Global Climate Change." *American Psychologist* 66, no. 4: 241–50.

Thompson, C. 2007. "Global Mourning: How the Next Victim of Climate Change Will Be Our Minds." *Wired* magazine. http://www.wired.com/techbiz/people/magazine/16-01/st_thompson. Accessed 5 August 2013.
Toffler, Alvin. 1970. *Future Shock*. London: The Bodley Head.
– 1975. *The Eco-Spasm Report*. New York: Bantam Books.
Tschakert, Petra, and Raymond Tutu. 2010. "Solastalgia: Environmentally Induced Distress and Migration among Africa's Poor Due to Climate Change." In *Environment, Forced Migration, and Social Vulnerability*, edited by T. Afifi and J. Jäger, 57–69. New York: Springer.
Tuan, Yi-Fu. 1974. *Topophilia: A Study of Environmental Perception, Attitudes, and Values*. New Jersey: Prentice-Hall, Inc.
Walker, B.H., and D. Salt. 2006. *Resilience Thinking: Sustaining Ecosystems and People in a Changing World.* Washington: Island Press.
Windle, Phyllis. 1992. "The Ecology of Grief." *Bioscience* 42, no. 5: 363–6.

# Epilogue

## The Wild Creatures

PATRICK LANE

*On 13 November 2013, the University of Victoria conferred an honorary degree on Patrick Lane. His convocation address, "An Open Letter to All the Wild Creatures of the Earth," published by the* Times Colonist *newspaper, has since been shared many times on other sites and through social media.*

*We thought this a fitting conclusion to a collection premised on loving and mourning wild creatures – ending with a call for hope, a call to action, and a call to looking toward the more-than-humans for their wisdom and for the lessons they can teach us about living in this world.*

It is sixty-five years ago, you're ten years old and sitting on an old, half-blind grey horse.

All you have is a saddle blanket and a rope for reins as you watch a pack of dogs rage at the foot of a Ponderosa pine. High up on a branch a cougar lies supine, one paw lazily swatting at the air. He knows the dogs will tire. They will slink away and then the cougar will climb down and go on with its life in the Blue Bush country south of Kamloops.

It is a hot summer day. There is the smell of pine needles and Oregon grape and dust. It seems to you that the sun carves the dust from the face of the broken rocks, carves and lifts it into the air where it mixes with the sun.

Just beyond you are three men on horses. The men have saddles and boots and rifles and their horses shy at the clamour of the dogs. The man with the Winchester rifle is the one who owns the dog pack and he is the one who has led you out of the valley, following the dogs through the hills to the big tree where the cougar is trapped. You watch as the man with the rifle climbs down from the saddle and sets his boots among the slippery pine needles. When the man is sure of his footing he lifts the rifle, takes aim, and then … and then you shrink inside a cowl of silence as the cougar falls.

As you watch, the men raise their rifles and shoot them at the sun. You will not understand their triumph, their exultance. Not then. You are too young. It will take years for you to understand. But one day you will step up to a podium in an auditorium at a university on an island far to the west and you will talk about what those men did.

You know now they shot at the sun because they wanted to bring a darkness into the world. Knowing that has changed you forever. Today I look back at their generation. Most of them are dead. They were born into the First Great War of the last century. Most of their fathers did not come home from the slaughter. Most of their mothers were left lost and lonely. Their youth was wasted through the years of the Great Depression when they wandered the country in search of work, a bed or blanket, a friendly hand, a woman's touch, a child's quick cry.

And then came the Second World War and more were lost. Millions upon millions of men, women, and children died in that old world. But sometimes we forget that untold numbers of creatures died with them: the sparrow and the rabbit, the salmon and the whale, the beetle and the butterfly, the deer and the wolf. And trees died too, the fir and spruce, the cedar and hemlock. Whole forests were sacrificed to the wars.

Those men bequeathed to me a devastated world.

When my generation came of age in the mid-century we were ready for change. And we tried to make it happen, but the ones who wanted change were few.

In the end we did what the generations before us did. We began to eat the world. We devoured the oceans and we devoured the land. We drank the lakes and the seas and we ate the mountains and the plains. We ate and we ate until there was almost nothing left for you or for your children to come.

The cougar that died that day back in 1949 was a question spoken into my life and I have tried to answer that question with my teaching, my poems, and my stories. Ten years after they killed the cougar I came of age. I had no education beyond high school, but I had a deep desire to become an artist, a poet. The death of the cougar stayed with me through the years of my young manhood.

Then, one moonlit night in 1963, I stepped out of my little trailer perched on the side of a mountain above the North Thompson River. Below me was the sawmill where I worked as a first-aid man. Down a short path a little creek purled through the trees just beyond my door. I went there under the moon and kneeling in the moss cupped water in my hands for a drink. As I looked up I saw a cougar leaning over his paws in the thin shadows. He was six feet away, drinking from the same pool. I stared at the cougar and found myself alive in the eyes of the great cat.

The cougar those men had killed when I was a boy came back to me. It was then I swore I would spend my life bearing witness to the past and the years to come.

I stand here looking out over this assembly and I ask myself what can I offer you who are taking from my generation's hand a troubled world. I am an elder now. There are times many of us old ones feel a deep regret, a profound sorrow, but our sorrow does not have to be yours. You are young and it is soon to be your time.

A month ago I sat on a river estuary in the Great Bear Rain Forest north of here as a mother grizzly nursed her cubs. As the little ones suckled, the milk spilled down her chest and belly. As I watched her I thought of this day and I thought of you who not so long ago nursed at your mother's breast. There in the last intact rain forest on earth, the bear cubs became emblems of hope to me.

Out there are men and women only a few years older than you who are trying to remedy a broken world. I know and respect their passion. You too can change things. Just remember there are people who will try to stop you and when they do you will have to fight for your lives and the lives of children to come. Today you are graduating with the degrees you have worked so hard to attain. They will affect your lives forever.

You are also one of the wild creatures of the earth. I want you for one moment to imagine you are a ten-year-old on a half-blind, grey horse. You are watching a cougar fall from the high limb of a ponderosa pine into a moil of raging dogs. The ones who have done this, the ones who have brought you here, are shooting at the sun. They are trying to bring a darkness into the world.

It's your story now. How do you want it to end?

# Contributors

GLENN ALBRECHT retired as professor of sustainability at Murdoch University in Perth, Western Australia, in June 2014. He was at the University of Newcastle as associate professor of environmental studies until December 2008. He is an environmental philosopher with both theoretical and applied interests in the relationship between ecosystem and human health, broadly defined. He pioneered the research domain of "psychoterratic" or earth-related mental health and emotional conditions with his concept of "solastalgia" or the lived experience of negative environmental change. Solastalgia has become accepted worldwide as a key concept in understanding the impact of negative environmental change in academic, creative arts, social impact assessment, and legal contexts. Albrecht's work is increasingly cited in scholarly literature and is now being used extensively in course readings, new research theses, and academic research in many disciplines including geography and environmental studies. His work is published in languages other than English. He has publications in the field of animal ethics, most recently on the ethics of relocating endangered species in the face of climate change pressures and the ethics of the thoroughbred horse industry worldwide. With Professor Phillip McManus (Sydney University) he has completed a book, published in 2012 by Routledge, on the thoroughbred industry. He also published with Professor McManus on the newly emerging domain of "psychoterratic geographies."

JESSICA MARION BARR is an artist and educator who received her PhD in cultural studies at Queen's University (Kingston, ON) in fall 2015. Her interdisciplinary practice incorporates artmaking, research, and pedagogy, focusing on ecological elegies. Jessica teaches courses, gives talks, and leads workshops on ecologically engaged art. Images of her artwork and accompanying texts have been published in *The Brock Review* and *The Goose*. She has recently exhibited work at the Gladstone Hotel

(Toronto); World of Threads Festival (Oakville); Wall Space Gallery (Ottawa); FINA Gallery, University of British Columbia (Kelowna); Hart House (University of Toronto); and Union Gallery, The Artel, Modern Fuel Artist-Run Centre, and the Ban Righ Centre (Kingston). Jessica's 2013 Nuit Blanche (Toronto) project *Indicator* was included in NOW magazine's Critics' Picks. *Vernal Pool*, a collaboration with artist Karen Abel, was the recipient of the 2014 Jury's Choice Award and the Ontario Association of Landscape Architects/GROUND Award at *Grow Op: Exploring Landscape + Place* at the Gladstone Hotel in Toronto (more on this participatory project at http://vernal-pool.tumblr.com); you can listen to Jessica's sound installation *Vernal Chorus* at https://soundcloud.com/jessica-marion-barr/vernal-chorus. Jessica is a member of the Association for Literature, Environment, and Culture in Canada (ALECC).

SEBASTIAN FELIX BRAUN studied ethnology, history, and philosophy at Universität Basel before earning his PhD in anthropology from Indiana University. He was an associate professor and chair of the department of American Indian Studies at the University of North Dakota, and is now associate professor of anthropology and director of American Indian Studies at Iowa State University. His interests have roots in kinship, worldview, language, and environment, and lie in the intersections of culture, politics, ecologies, and economics. Braun is the author of *Buffalo Inc.: American Indians and Economic Development* (University of Oklahoma Press, 2008) and editor of *Transforming Ethnohistories: Narrative, Meaning, and Community* (University of Oklahoma Press, 2013). Among other things, he has written for the International Work Group for Indigenous Affairs' (IWGIA) *The Indigenous World* since 2004. His current focus is on extraction of energy resources and energy booms.

ASHLEE CUNSOLO is a passionate researcher, environmental advocate, and ally working with research and policy to make a difference in how we live with and in this world. As a community-engaged social science and health researcher working at the intersection of place, culture, health, and environment she has spent a decade working with Indigenous communities and leaders across Canada on a variety of community-led and community-identified research initiatives ranging from the impact of climate change on physical and mental health to cultural reclamation and intergenerational knowledge transmission, suicide reduction and prevention, land-based education and healing programs, environmental grief and mourning, and Indigenization of higher education. She is a pioneer in climate change and mental health research, and has given over 200 talks and received wide media attention for her work. In 2014 she released a documentary film, collaboratively produced with the five Inuit communities in Nunatsiavut, Labrador, about the impacts of climate change on Inuit culture, livelihoods, and well-being (www.lamentfortheland.ca). She is currently the director of the Labrador Institute of Memorial University, an inaugural member of the Royal Society of Canada's College of New Scholars, Artists, and Scientists, and one of the founding members of Nature Canada's 75 Women for Nature.

AMANDA DI BATTISTA is a PhD candidate in the Faculty of Environmental Studies at York University, where she explores the importance of creative writing and literature for environmental education and ecopolitics. She is co-producer of *CoHearence,* a podcast series that examines the relationship between culture and the environment, as well as an editor with *UnderCurrents: Journal of Critical Environmental Studies* and the reviews editor for *The Goose*, the Association for Literature, Environment, and Culture in Canada's (ALECC) online publication.

FRANKLIN GINN is a lecturer in human geography at the University of Edinburgh, where his research focuses on more-than-human geographies. He has published on domestic garden cultures and is currently completing a book, *Domestic Wild.* His current research explores cultures of apocalypse and the Anthropocene and, as part of an AHRC-funded project, religious responses to climate ethics.

BERNIE KRAUSE has travelled the world recording and archiving the sounds of creatures and environments large and small since 1968. Working at the research sites of Jane Goodall and Dian Fossey, he identified the concept of *biophony* based on the relationships of individual creatures to the total biological soundscape. His contributions helped establish the foundation of soundscape ecology. Krause has produced over fifty natural soundscape CDs, in addition to the design of interactive, non-redundant environmental sound sculptures for museums and public spaces around the world. Krause's new book, *The Great Animal Orchestra: Finding the Origins of Music in the World's Wild Places*, was published by Little Brown in 2012, and has been translated into seven languages. In July 2014 the Cheltenham Music Festival premiered *The Great Animal Orchestra: Symphony for Orchestra and Wild Soundscapes*, a new symphony by Oxford composer Richard Blackford, with biophonies by Krause and featuring the BBC National Orchestra of Wales. Previously Krause was a professional studio musician, and his work has appeared on over 250 albums including those of Van Morrison, Brian Eno and David Byrne, George Harrison, and the Doors. Krause's music has also been used in 135 films including *Apocalypse Now*, *Rosemary's Baby*, and *Castaway*.

LISA KRETZ is an assistant professor of philosophy at the University of Evansville. Her recent and forthcoming publications are at the intersection of areas such as ecological selfhood, the theory-action gap, motivational framing for action, ecological emotions, moral psychology, the politics of emotion, and activist pedagogy. She also has a number of publications that identify and argue against the immoral treatment of non-human animals, and she has been vegan for the last twelve years. Her work can be found in journals such as the *Journal of Agricultural and Environmental Ethics*, *Ethics & the Environment*, the *Journal for Critical Animal Studies*, and *Environmental Ethics: An Interdisciplinary Journal Dedicated to the Philosophical Aspects of Environmental Problems*.

KAREN LANDMAN is a professor of landscape architecture in the School of Environmental Design and Rural Development at the University of Guelph. Her background is in horticulture, landscape architecture, landscape planning, and cultural geography. Karen's current research interests include landscape stewardship, green infrastructure, pollination habitat, and urban agriculture as a form of land stewardship. With a team of graduate students and conservation professionals, Landman has developed the Ontario Rural Landowner Stewardship Guide to assist landowners who wish to steward natural features and environmental quality on their properties, thereby contributing to the long-term protection of the greater landscape.

PATRICK LANE is one of Canada's pre-eminent poets, an Officer of the Order of Canada, and a winner of numerous awards including the Governor General's Award for Poetry, the Canadian Authors Association Award, the Lieutenant Governor's Award for Literary Excellence, and three National Magazine Awards. His distinguished career spans fifty years and twenty-five volumes of poetry as well as award-winning books of fiction and non-fiction. He has been a writer in residence and teacher at Concordia University in Montreal, the University of Victoria, and the University of Toronto. Patrick Lane lives near Victoria with his wife, the poet Lorna Crozier.

ANDREW MARK has a BA from the McGill School of Environment and a PhD from the Faculty of Environmental Studies at York University, where he also holds an MA in ethnomusicology. His research interests concern the importance of the arts for ecological governance in community, dialogues of idealism, and the relationship between making music and environmental ethos. His dissertation considers the ecomusicology of Hornby Island in British Columbia and the social and environmental stresses that the small community endures to replicate itself. Andrew is a co-producer of the podcast series *CoHearence*, is a board member of the *Ecomusicology Newsletter* and *Undercurrents: The Journal of Critical Environmental Studies*, has published in *Environmental Humanities*, *Ethnomusicology Review*, *Ethnologies*, and has work forthcoming in several edited books. He is a professional musician, sound artist, and parent.

NANCY MENNING is assistant professor of world religions at Ithaca College in upstate New York. She completed her PhD in 2010 from the Department of Religious Studies at the University of Iowa. Her dissertation, "Reading Nature Religiously: Lectio Divina, Environmental Ethics, and the Literary Nonfiction of Terry Tempest Williams," described and illustrated a mode of constructing environmental ethics grounded in vulnerability and intimacy. Dr Menning combines her background in social forestry and environmental studies with her work in religious studies to advance the environmental humanities and to foster simultaneous human and ecological flourishing. Her recent work has emphasized religious aspects of climate change narratives, spiritual practices for cultivating environmental virtues, rituals for mourning environmental losses, and pedagogical strategies for teaching religion and nature using literature. She serves as co-chair of the Ecology and Science in the Study of Religion section of the Midwest region of the American Academy of Religion.

JOHN CHARLES RYAN is postdoctoral research fellow in the School of Arts at the University of New England in Australia and honorary research fellow in the School of Humanities at the University of Western Australia. From 2012 to 2015, he was postdoctoral research fellow in the School of Communications and Arts at Edith Cowan University. His teaching and research activities cross between the environmental and digital humanities. In particular, he has contributed to the fields of ecocriticism (Australian and Southeast Asian) and critical plant studies. He is the author, co-author, editor, or co-editor of ten books, including *Digital Arts: An Introduction to New Media* (Bloomsbury, 2014) and *The Language of Plants: Science, Philosophy, Literature* (University of Minnesota Press, 2017).

CATRIONA SANDILANDS is a professor in the Faculty of Environmental Studies at York University. Her work lies at the intersections of queer and feminist theory, environmental and posthumanist thought and politics, and environmental literary criticism and cultural studies. Her publications include *The Good-Natured Feminist: Ecofeminism and Democracy*; *This Elusive Land: Women and the Canadian Environment* (with Melody Hessing and Rebecca Raglon); *Queer Ecologies: Sex, Nature, Politics, Desire* (with Bruce Erickson); and, forthcoming, *Green Words/Green Worlds: Environmental Literatures and Politics* (with Ella Soper and Amanda Di Battista). Recent works also include invited chapters in the *Oxford Handbook of Ecocriticism*; the *Cambridge Companion to Literature and Environment*; the *Oxford Handbook of Environmental Political Theory*; and, forthcoming, the *Routledge International Handbook on Gender and Environment*. Her current research involves two projects, the first a critical treatment of Canadian novelist, essayist, and activist Jane Rule and the multiple publics of her public intellectualism, and the second an exploration, in literary non-fiction, of urban plant cultures in Toronto, *Plantasmagoria*.

HELEN WHALE completed an MSc in environment, culture, and society at the University of Edinburgh in 2012, where fieldwork into the disappearance and conservation of the house sparrow formed the topic for her research dissertation. Since graduating she has held two roles in people engagement and education outreach for the Royal Society for the Protection of Birds, Europe's largest conservation organization. She recently lived on Rathlin Island, off the north coast of Northern Ireland, where she was involved with a pilot project to sustainably cultivate, harvest, and process kelp seaweeds for the local and international food markets. Helen has recently moved from Rathlin Island back to her native Yorkshire, where she continues to work within the charity sector.

# Index